Operational Amplifier Circuits: Analysis and Design

Operation Amplifier Circuits: Analysis and Design

John C.C. Nelson

Butterworth–Heinemann
Boston Oxford Melbourne Singapore Toronto Munich New Delhi Tokyo

Recognizing the importance of preserving what has been written, Butterworth–Heinemann prints its books on acid-free paper whenever possible.

Library of Congress Cataloging-in-Publication Data

Nelson, J.C.C. (John Christopher Cunliffe), 1938–
Operational amplifier circuits: analysis and design / by John C.C. Nelson
p. cm.
Includes bibliographical references and index.
ISBN 0-7506-9468-8
1. Operational amplifiers—Design and construction. I. Title.
TK78671.58.06N454 1994
621.39'5—dc20 94-32724
CIP

British Library Cataloguing-in-Publication Data

A catalogue record for this book is available from the British Library.

Butterworth–Heinemann
313 Washington Street
Newton, MA, 02158-1626

10 9 8 7 6 5 4 3 2 1

Printed in the United States of America

To Sue, Joanna, and Samantha

Contents

Preface

The operational amplifier is essentially an electronic circuit capable of producing an output that is related to its input by a known mathematical operation. Originally such circuits were cumbersome and expensive, since they made use of several thermionic vacuum tubes and, subsequently, discrete transistors. Today "op amps," as they have become known popularly, are available as integrated circuit "chips" at very low cost. Four chips, costing twenty cents or less each, can be accommodated in one small package. Consequently chips are used in a remarkably wide range of applications, not all of which are directly related to the original intention of performing mathematical operations. Most of the important application areas are discussed in this book.

All electronic circuit design involves substantial calculation in order to meet the required specification. One of the advantages of operational amplifier circuits is that the assumptions of ideal operation which are normally made (see sections 1.1 and 1.2, pages 1 and 3) often lead to relatively simple design equations for which a pocket calculator is quite adequate.

However, some of the calculations—particularly those where several attempts are necessary in order to obtain the required performance with readily available component values—justify the use of a computer. In other cases, particularly the behavior of circuits with respect to frequency, a computer-generated graphical display can be the most convenient way to assess predicted performance.

For these reasons, the text, which is an updated version of the author's *BASIC Operational Amplifiers* (Butterworth, 1986), is illustrated with a range of computer programs (see Appendix A, page 107) which may be used for serious circuit design and also to examine the effects of a wide range of parameter values in order to illustrate points made in the text. The Pascal language was chosen because of its excellent structuring and because its code is virtually self-documenting.

This book assumes a background in the basic techniques of circuit analysis—particularly the use of j notation for reactive circuits—with a corresponding level of mathematical ability. The Laplace transform is used in the chapter on active filters

(Chapter 5, page 57) but not elsewhere. Practical considerations in the use of operational amplifiers are not discussed in detail; for this the reader is referred to a practically oriented text. Many are available, and Jung's *IC Op-Amp Cookbook* (Howard W. Sams, 1986) has become a bible in this context. It is referenced throughout the text wherever practical aspects are important.

The author gratefully acknowledges the valued suggestions made by Robert Craven of Teradyne Inc.; the highly detailed comments and helpful assistance of Edwin Richter, the series editor; and the patient support provided by his wife, Sue, during the long manuscript editing period.

CHAPTER 1

Introduction to Operational Amplifier Circuits

1.1 The Basic Amplifier

The basic amplifier may be represented by the symbol shown in Figure 1–1. The amplifier has two inputs, which are denoted by V_{i+} and V_{i-}, and a single output, V_o. Positive and negative power supplies of equal magnitude are normally used (although single-supply operation is possible) and are shown as $+V_s$ and $-V_s$ in Figure 1–1 (for simplicity these connections are not normally shown

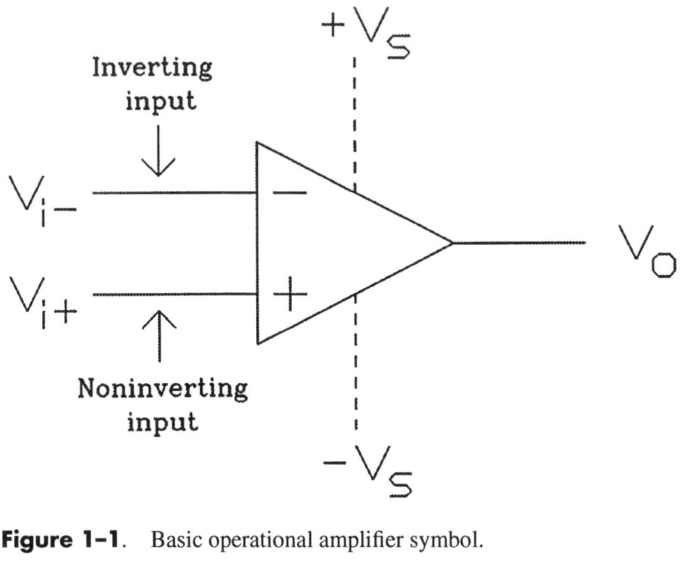

Figure 1–1. Basic operational amplifier symbol.

on circuit diagrams). The common zero of $+V_s$ and $-V_s$ is an important reference value for V_{i+}, V_{i-}, and V_o that does not appear explicitly on the amplifier symbol, since a direct connection is not required. However, one of the amplifier inputs may be connected to it either directly or indirectly, depending on the required mode of operation.

Ideal operation of the amplifier is shown in the transfer characteristic of Figure 1–2. Here V_i represents the difference between the voltages applied to the two inputs (V_{i+} and V_{i-}). It can be seen that if V_i is positive, even by only a small amount, the output V_o is positive and constant, having a magnitude slightly less than that of the supply voltage (the *output saturation voltage*). Similarly, negative values of V_i produce a constant negative output.

In practice, a finite change in V_i will be needed in order to change V_o from one level to the other, as shown by the dotted line in Figure 1–2. Also, the changeover will occur for a value of V_i that is not precisely equal to 0. (This effect will be discussed further in Chapter 3.)

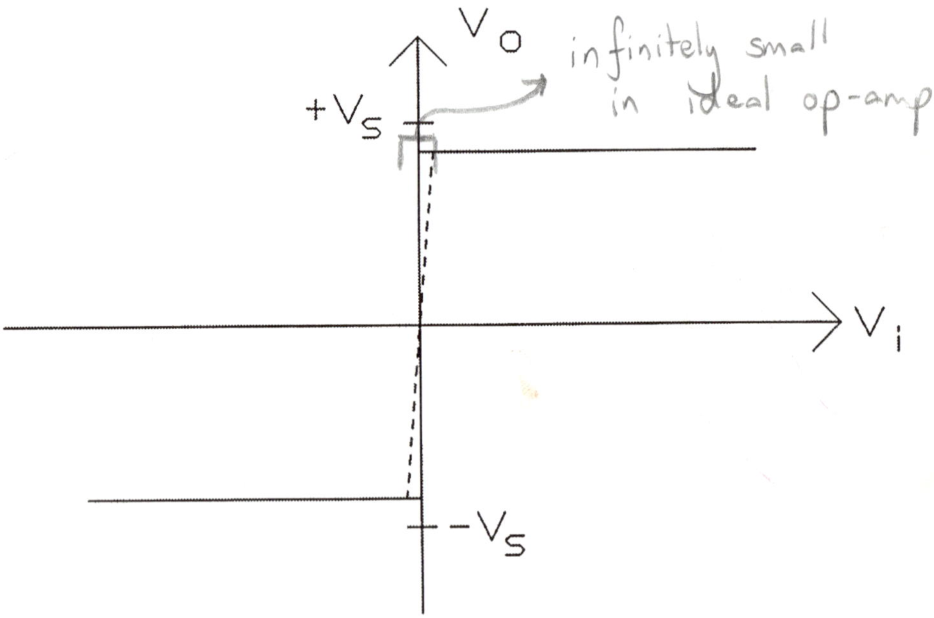

Figure 1–2. Ideal transfer characteristic (solid line) and practical approximation (broken line) ($V_i = V_{i+} - V_{i-}$).

For a characteristic having a finite slope, the input/output relationship may be written as

$$V_o = A\,(V_{i+} - V_{i-}), \tag{1.1}$$

where A is the gain of the amplifier in the region between the two output saturation voltages. The value of A is large for practical amplifiers (typically more than 50,000) and theoretically infinite for ideal ones. A is known as the *open loop gain,* which is the gain of the amplifier without feedback (an external connection that makes V_i depend on V_o in some way). The inputs (indicated by + and – in Figure 1–1) are referred to as *noninverting* and *inverting,* respectively, for reasons that are evident from Equation (1.1).

The amplifier can be used in the basic form described above in order to distinguish between positive and negative input values. If used in this manner it would be described as a *comparator,* and the output levels would normally be constrained to levels suitable for connection to digital logic circuits. An application of a comparator will be discussed briefly in Section 4.2 (page 45). In the present context, a continuous relationship between input and output is required and is achieved by means of feedback. Several different configurations are widely used and are discussed in the following sections. Operation without feedback is often referred to as *open loop operation,* which becomes *closed loop operation* when feedback is applied; that is, when the feedback loop is closed.

1.2 Inverting Mode, Operation as Scaler and Summer

The basic configuration is shown in Figure 1–3, where the resistors R_i and R_f are the input and feedback resistors, respectively. The noninverting input of the amplifier is connected to the common zero of the power supplies (shown as a chassis, or ground, connection in Figure 1–3), and the inverting input has a voltage v with respect to this. Let the currents in the input and feedback resistors be i_i and i_f, as shown. If the input resistance of the amplifier itself is so high that the current flowing into the inverting input may be neglected—an assumption that is normally justified in practice— the currents will sum to 0: $i_i + i_f = 0$. Ohm's law can be applied to each resistor:

$$\frac{(V_i - v)}{R_i} + \frac{(V_o - v)}{R_f} = 0. \tag{1.2}$$

In this configuration, Equation (1.1) becomes $V_o = -Av$ and therefore,

$$v = -V_o/A \tag{1.3}$$

Substituting this into (1.2) yields:

$$\frac{V_i + \frac{V_o}{A}}{R_i} + \frac{V_o + \frac{V_o}{A}}{R_f} = 0. \tag{1.4}$$

For large values of *A*, *v* tends toward 0 and this reduces to

$$\frac{V_i}{R_i} + \frac{V_o}{R_f} = 0 \tag{1.5}$$

or

$$V_o = -\left(\frac{R_f}{R_i}\right)V_i. \tag{1.6}$$

This is an important and useful result since the relationship between V_o and V_i (a "gain" of $-R_f/R_i$) depends only on the values of the resistors and not on the characteristics of the amplifier itself. This is true, of course, only when the circuit is operating under such conditions that the assumptions of negligible amplifier input current and very high open loop gain are valid. Since *v* has become very small, the potential of the inverting input is very close to that of the common reference. Consequently, this point is often referred to as a *virtual ground*.

The circuit of Figure 1–3 is, therefore, capable of multiplying the input voltage by a negative constant that may be made less than, equal to, or greater than 1 by an appropriate choice of R_f and R_i. This process is often described as *scaling*. A straightforward extension to this circuit allows several input voltages to be added and scaled if required, as shown in Figure 1–4. Summing the input and feedback currents as before yields:

$$V_o = -\left(\frac{R_f}{R_{i1}}V_{i1} + \frac{R_f}{R_{i2}}V_{i2} + \frac{R_f}{R_{i3}}V_{i3} + \frac{R_f}{R_{i4}}V_{i4}\right). \tag{1.7}$$

Notice that a change in R_f alters the scaling of all the inputs, and each of the input resistance values can be used to define the scaling of the individual

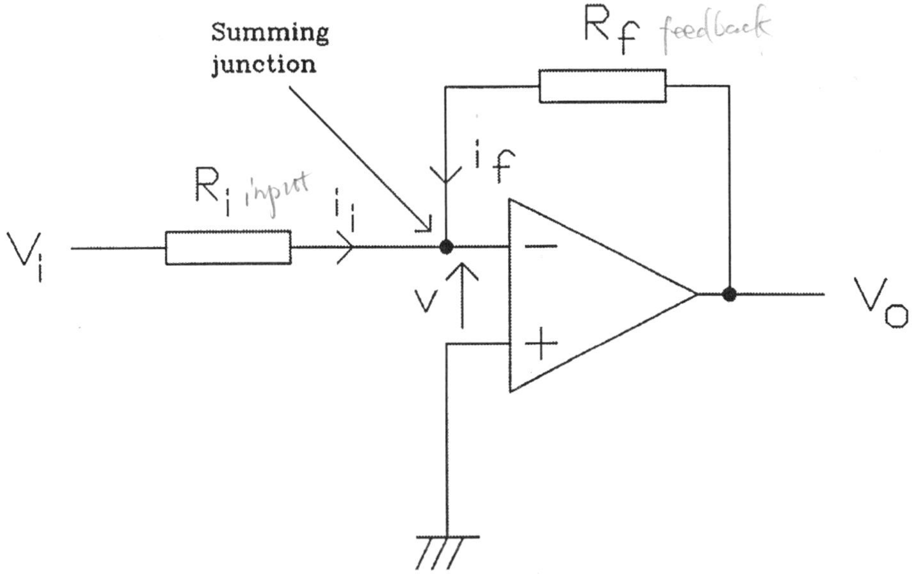

Figure 1–3. Operational amplifier configuration.

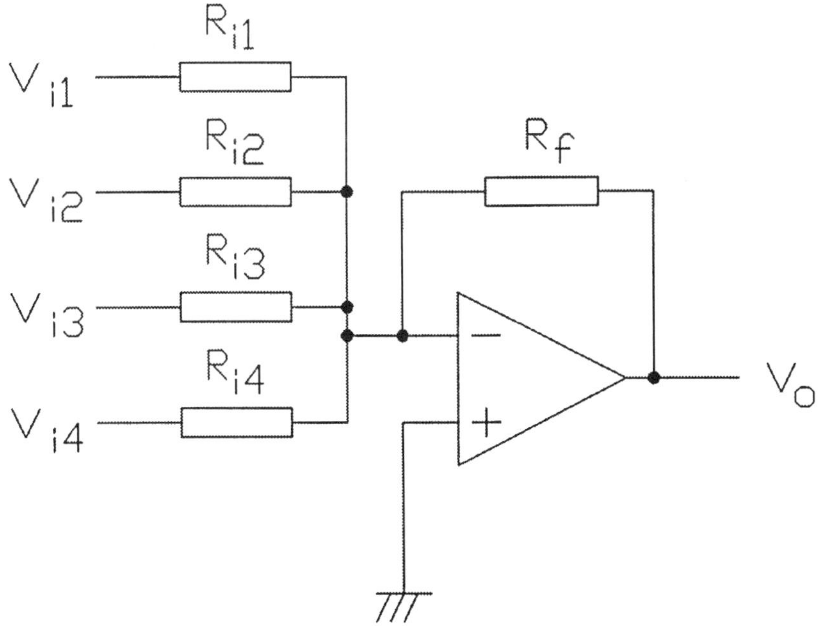

Figure 1–4. Circuit for summing several inputs.

inputs. The number of inputs is not limited to four, of course, but a practical limit is imposed by the fact that the sum of all the input currents must be balanced by the amplifier output current flowing through R_f. The junction of the input and feedback resistors at the noninverting input is often referred to as the *summing junction* (it is also the virtual ground point).

As an example, consider generation of an output voltage V_o such that

$$V_o = -(V_{i1} + 2V_{i2} + 3V_{i3} + 4V_{i4}) . \tag{1.8}$$

Comparison with Equation (1.7) shows that

$$\frac{R_f}{R_{i1}} = 1, \ \frac{R_f}{R_{i2}} = 2, \ \frac{R_f}{R_{i3}} = 3, \text{ and } \frac{R_f}{R_{i4}} = 4.$$

Since these scaling factors are defined by the ratio of two resistances, an arbitrary choice of one value must be made. A useful intermediate step is to specify all resistance values as multiples of a basic value R. It is convenient to let the input resistance for the *highest* required gain be R; so in this case $R_{i4} = R$. Hence $R_f = 4R$, $R_{i3} = R_f/3 = 4R/3$, $R_{i2} = R_f/2 = 2R$, and $R_{i1} = R_f = 4R$, as shown in Figure 1–5.

Although the scaling factors of Equation (1.7) may be chosen independently for each of the inputs, they are all of the same sign (negative for a single inverting amplifier). Both positive and negative scaling factors may be accommodated by adding a second amplifier, as shown in Figure 1–6.

Applying Equation (1.7) to amplifier 1,

$$V_{o1} = -\left(V_{i1}\frac{R_{f1}}{R_{i1}} + V_{i2}\frac{R_{f1}}{R_{i2}}\right). \tag{1.9}$$

Similarly, for amplifier 2,

$$V_{o2} = -\left(V_{i3}\frac{R_{f2}}{R_{i3}} + V_{i4}\frac{R_{f2}}{R_{i4}} + V_{o1}\frac{R_{f2}}{R_{i5}}\right). \tag{1.10}$$

Substituting (1.9) into (1.10) to eliminate V_{o1},

$$V_o = V_{o2} = -\left[V_{i3}\frac{R_{f2}}{R_{i3}} + V_{i4}\frac{R_{f2}}{R_{i4}} - \left(V_{i1}\frac{R_{f1}}{R_{i1}} + V_{i2}\frac{R_{f1}}{R_{i2}}\right)\frac{R_{f2}}{R_{i5}}\right]. \tag{1.11}$$

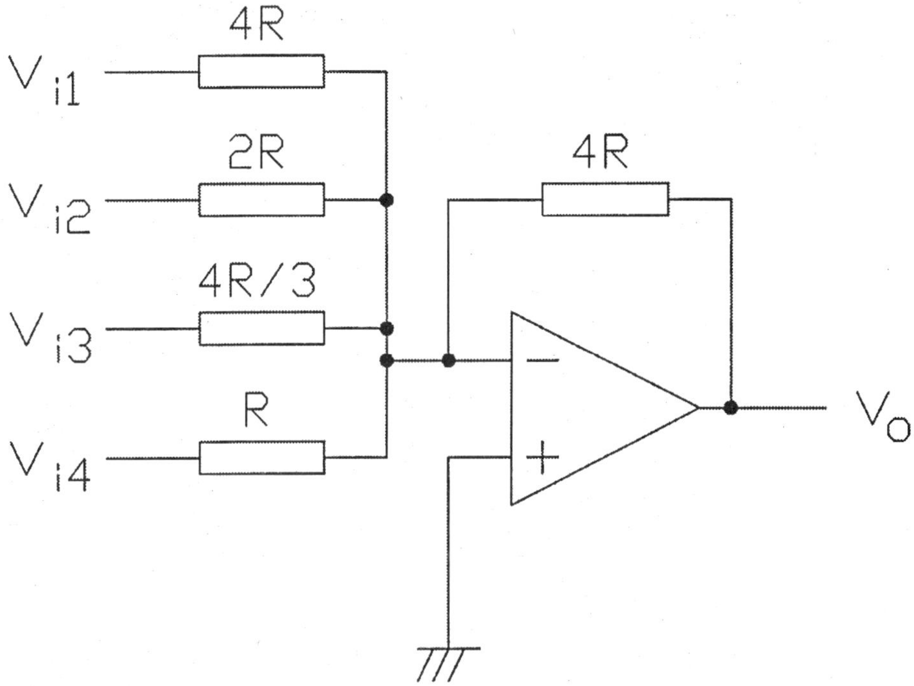

Figure 1–5. Operational amplifier circuit to obtain $V_o = -[V_{i1} + 2V_{i2} + 3V_{i3} + 4V_{i4}]$.

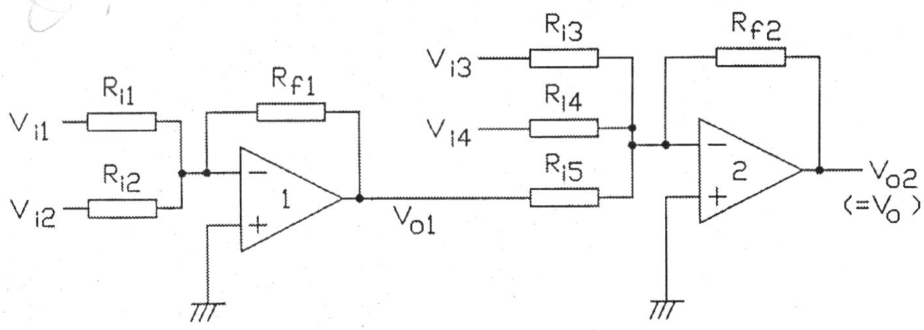

Figure 1–6. Two operational amplifiers used to obtain coefficients of either sign.

Rearranging terms and letting $R_{i5} = R_{f2}$ to simplify the scaling factors,

$$V_o = V_{i1}\frac{R_{f1}}{R_{i1}} + V_{i2}\frac{R_{f1}}{R_{i2}} - V_{i3}\frac{R_{f2}}{R_{i3}} - V_{i4}\frac{R_{f2}}{R_{i4}}. \tag{1.12}$$

From this it can be seen that inputs V_{i1} and V_{i2}, which pass through both amplifiers, have a positive scaling factor, while V_{i3} and V_{i4}, which pass through amplifier 2 only, have negative scaling factors.

As an example, consider generation of $V_o = V_{i1} + 3V_{i2} - 2V_{i3} - 4V_{i4}$ using the approach described above. Let $R_{i4} = R$; hence $R_{f2} = 4R$ and $R_{i3} = R_{f2}/2 = 2R$. Similarly for amplifier 1, let $R_{i2} = R$, $R_{f1} = 3R$, and $R_{i1} = R_{f1} = 3R$. The corresponding circuit is shown in Figure 1–7.

Input and feedback resistance values have so far been specified as a multiple of the basic value R. However, there are practical limits to the values that may be employed, particularly for the feedback resistance. For Equations (1.6) and (1.7) to hold under all conditions, the amplifier must be capable of providing an output current that exceeds V_o/R_f when V_o is at its maximum (saturation) value (see Figures 1–2 and 1–3). This imposes a minimum value on R_f; attempts to use a lower value cause the summing junction to be "pulled" away from its value close to 0 and the circuit ceases to operate correctly. For simplicity, the maximum output voltage can be taken to be the supply voltage (it will normally be slightly less than this in practice).

For example, if the amplifier can provide a maximum output of 10 mA with a 15 V supply, a feedback resistance R_f of 1.5 kΩ (calculated as 15 V ÷ 10 mA) would consume all of the available current, leaving none to provide a useful output from the circuit. A lower value, such as 1.0 kΩ, would drop only 10 V at 10 mA, the virtual ground would be at +5 V, and Equations (1.6) and (1.7) would no longer be valid. The maximum amplifier output voltage divided by the maximum output current (1.5 kΩ in this example), therefore, provides an absolute minimum value for the feedback resistance, and a practical circuit should use a value significantly greater than this.

At the other extreme, very high values of feedback resistance (typically in excess of 1 MΩ) should normally be avoided, since the offset error due to

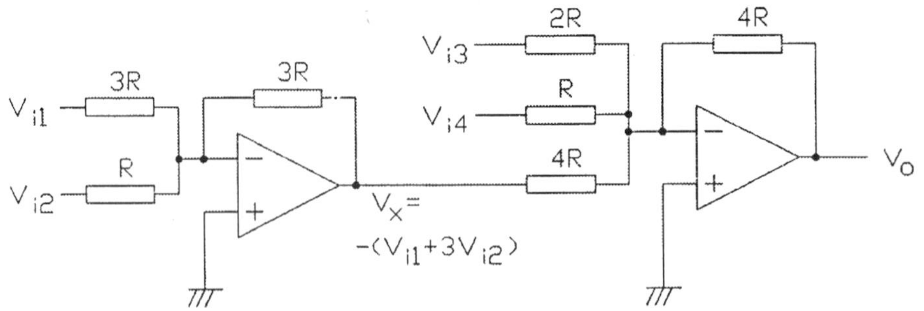

Figure 1–7. Realization of $V_{i1} + 3V_{i2} - 2V_{i3} - 4V_{i4}$ using two operational amplifiers.

bias current (see Section 3.1, page 35) can become large. Equation (3.8) in Chapter 3 shows that the offset error due to input bias current is directly proportional to the product of the feedback resistance itself and the bias current. For example, an amplifier with an input bias current of 100 nA and a feedback resistance of 10 kΩ would produce an offset error of 1 mV, whereas a feedback resistance of 1 MΩ would produce an error of 0.1 V for the same bias current.

The output resistance of operational amplifiers is very low, since an already low value (a few tens of ohms) for the basic amplifier is reduced by a factor of the order of the open loop gain A in the operational amplifier configuration. This means that the amplifier output closely resembles an ideal voltage source (the voltage is not affected by any reasonable load that may be connected). The magnitude of this voltage depends only on the input(s) and associated passive components—a *controlled voltage source.* Amplifiers of this kind can, therefore, be connected in cascade without significant interaction. This in turn means that complex functions may be realized by the interconnection of "building blocks" that perform the required basic operations.

Note also, however, that determination of the minimum practical value of feedback resistance discussed above must also take account of all other loads to be driven. Since the effective input resistance of an operational amplifier at the summing junction is very low, the apparent input resistance at the remote end of the input resistor is very nearly equal to R_i itself. The total load to be driven is, therefore, the parallel combination of the feedback resistance R_f and the input resistance R_i for each driven amplifier together with any additional loads. Let the parallel combination of input resistances and external loads be denoted by R_L. The overall value of the total load resistance must exceed V_S/I_{max} where V_S is the supply voltage and I_{max} is the maximum amplifier output current. Hence the limiting case is given by

$$\frac{R_{fmin}R_L}{R_{fmin} + R_L} = \frac{V_s}{I_{max}}, \tag{1.13}$$

where R_{fmin} is the minimum possible feedback resistance and the left-hand side of the equation represents R_{fmin} and R_L in parallel.

Therefore, by cross multiplying and extracting R_{fmin},

$$R_{fmin} = \frac{R_L V_s}{I_{max} R_L - V_s}. \tag{1.14}$$

Manufacturers do not necessarily quote a maximum output current capability in their data sheets; in many cases a minimum load resistance for a guaranteed output voltage swing is quoted. In any case, the output current is limited (typically to around 20 mA) to prevent damage to the amplifier if its output should become short-circuited (Jung 1986, 66). The computer program Input and Feedback Resistance Values (Inverting Amplifier) (Appendix A.1) facilitates calculation of the minimum feedback resistance value and determination of the input resistances required to achieve specified gains.

1.3 Noninverting Mode, the Voltage Follower

In some applications, the sign change associated with the inverting mode of operation is not required. A noninverting configuration is shown in Figure 1–8. The potential v_- at the inverting input may be derived from the output voltage V_o, since R_i and R_f form a potential divider:

$$v_- = V_o \frac{R_i}{R_f + R_i}. \tag{1.15}$$

The notation R_f and R_i has been retained, although R_i is not now directly associated with the input. As before, current flowing into the amplifier is assumed to be negligible.

From Equation (1.1), $V_o = A(V_i - v_-)$. Therefore,

$$V_i - v_- = \frac{V_o}{A}. \tag{1.16}$$

As A becomes very large, V_i becomes very nearly equal to v_- and, therefore, from Equation (1.15),

$$V_i = V_o \frac{R_i}{R_f + R_i}. \tag{1.17}$$

Therefore,

$$V_o = \frac{V_i(R_f + R_i)}{R_i} = V_i\left(1 + \frac{R_f}{R_i}\right). \tag{1.18}$$

As before, the relationship between input and output voltages depends only on R_f and R_i. This time the constant of proportionality is positive and of a

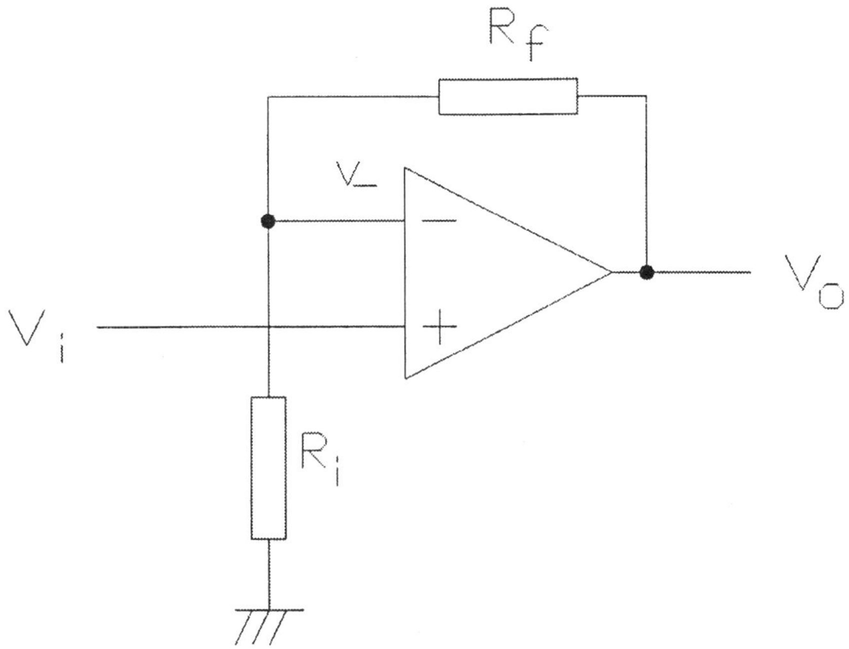

Figure 1–8. Noninverting feedback amplifier.

slightly different form (compare Equations [1.6] and [1.18]); notice that the gain $V_o/V_i = 1 + R_f/R_i$ cannot be less than unity.

Suitable component values (for applications other than unity gain and for the case of a single input) may be determined using the computer program Input and Feedback Resistance Values (Noninverting Amplifier) (Appendix A.2), which is an appropriately modified version of the program Input and Feedback Resistance Values (Inverting Amplifier) (Appendix A.1).

Several input resistors can be used to provide summation without inversion, as shown in Figure 1–9. However, the scaling factors are more complex and may be derived as follows: the voltage v_+ at the noninverting input is V_{i2} plus a fraction of the difference between V_{i1} and V_{i2} defined by the potential divider formed by R_{i1} and R_{i2} (current flow into the amplifier is neglected as usual), or

$$v_+ = V_{i2} + (V_{i1} - V_{i2}) \frac{R_{i2}}{R_{i1} + R_{i2}} \tag{1.19}$$

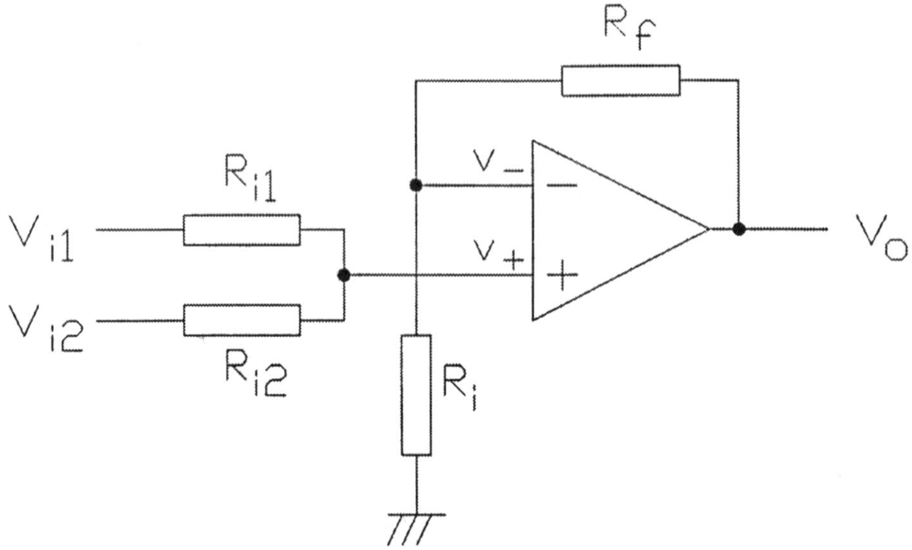

Figure 1–9. Noninverting summer.

$$\therefore\ v_+ = \frac{V_{i2}(R_{i1}+R_{i2}) + V_{i1}R_{i2} - V_{i2}R_{i2}}{R_{i1}+R_{i2}} = \frac{V_{i2}R_{i1}+V_{i1}R_{i2}}{R_{i1}+R_{i2}}. \tag{1.20}$$

The voltage v_- at the inverting input is again defined as in Equation (1.15):

$$v_- = V_o\frac{R_i}{R_i+R_f}. \tag{1.21}$$

If the normal assumption of very high open loop gain is made, $v_+ \approx v_-$ and the right-hand sides of Equations (1.20) and (1.21) can be equated:

$$V_o \bullet \frac{R_i}{R_i+R_f} = \frac{V_{i2}R_{i1}+V_{i1}R_{i2}}{R_{i1}+R_{i2}} \tag{1.22}$$

$$\therefore\ V_o = V_{i1}\left(\frac{R_{i2}}{R_{i1}+R_{i2}}\right)\left(1+\frac{R_f}{R_i}\right) + V_{i2}\left(\frac{R_{i1}}{R_{i1}+R_{i2}}\right)\left(1+\frac{R_f}{R_i}\right). \tag{1.23}$$

Since feedback action causes the potential of the inverting input to follow that of the noninverting input very closely, negligible current is drawn from the source and the circuit has a very high effective input resistance. This, combined with a very low output resistance as for the inverting configuration, makes this circuit a particularly useful one.

In particular, if R_f is made equal to 0 (for the case of a single input noninverting amplifier), the gain becomes equal to unity regardless of the value of R_i (Equation [1.18]) and R_i can be omitted. The result is a simple, but highly effective, unity gain buffer amplifier usually known as a *voltage follower* (see Figure 1–10). It is particularly suitable for preventing interaction between cascaded sections of a circuit, as outlined in Figure 1–11. In the figure, the output of the first section is represented as a voltage source V_{o1} in series with an effective output impedance Z_o (a Thévenin equivalent model) and the input impedance of the second section is represented by Z_i. In the absence of the voltage follower (with a direct connection between Z_o and Z_i), the output of the first section would be modified by the loading effect of Z_i (V_{o1} would be modified by the potential divider action of Z_i and Z_o). With the voltage follower connected as shown, the input impedance of the follower at the noninverting input is so high that V_i is very nearly equal to V_{o1} and, since feedback action forces

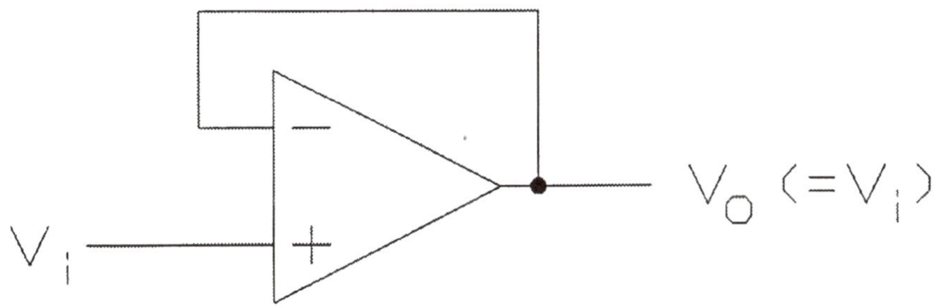

Figure 1–10. Unity gain buffer amplifier or voltage follower.

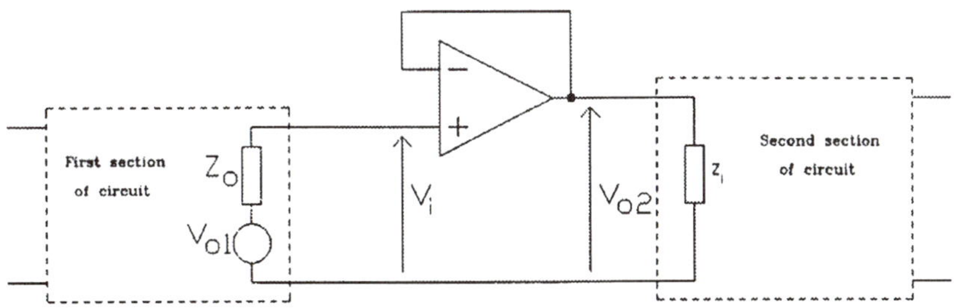

Figure 1–11. Voltage follower used to isolate two cascaded circuits.

the inverting input to the same value as the noninverting one, V_{o2} will also be very nearly equal to V_{o1}. As the output impedance of the voltage follower is very low, the input to the second section of the circuit will be very close to V_{o1} for all practical values of Z_i.

1.4 Differential Mode

The inverting and noninverting configurations, discussed in the previous sections, can be combined in order to obtain a difference signal using a single amplifier. A suitable circuit is shown in Figure 1–12.

Summing currents at the inverting input as before and letting the voltage at the inverting input be v_-,

$$\frac{V_{i1} - v_-}{R_{i1}} + \frac{V_o - v_-}{R_f} = 0. \tag{1.24}$$

By potential divider action, v_+, the voltage at the noninverting input, is given by

$$v_+ = V_{i2}\frac{R_g}{R_{i2} + R_g} = V_{i2}R', \tag{1.25}$$

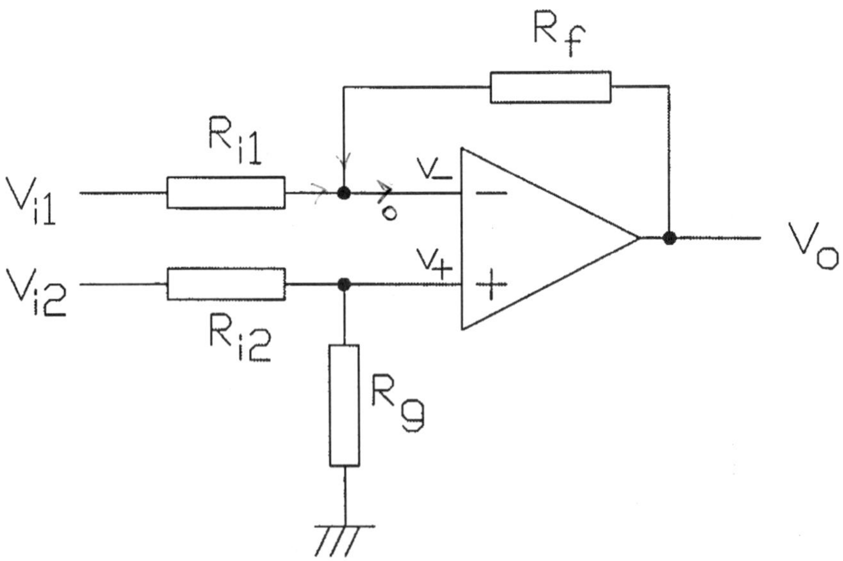

Figure 1–12. Difference, or differential, amplifier configuration.

where

$$R' = R_g / (R_{i2} + R_g). \tag{1.26}$$

(In both cases current flow into the amplifier is neglected, as usual.)

The high open loop gain of the amplifier ensures that v_+ will be very nearly equal to v_- and hence, substituting (1.25) into (1.24),

$$\frac{V_{i1} - V_{i2}R'}{R_{i1}} + \frac{V_o - V_{i2}R'}{R_f} = 0. \tag{1.27}$$

Therefore,

$$\frac{V_o}{R_f} = V_{i2}\left(\frac{R'}{R_{i1}} + \frac{R'}{R_f}\right) - \frac{V_{i1}}{R_{i1}}. \tag{1.28}$$

Substituting for R' using Equation (1.26),

$$\frac{V_o}{R_f} = V_{i2}\frac{R_g}{R_{i2} + R_g}\left(\frac{1}{R_{i1}} + \frac{1}{R_f}\right) - \frac{V_{i1}}{R_{i1}}. \tag{1.29}$$

Multiplying by R_f and rearranging,

$$V_o = V_{i2}\frac{R_g}{R_{i2} + R_g}\left(1 + \frac{R_f}{R_{i1}}\right) - V_{i1}\frac{R_f}{R_{i1}}. \tag{1.30}$$

Notice that the (negative) gain with respect to input V_{i1} can be varied over a wide range by choice of R_f and R_{i1} as before. However, for a given ratio of R_f and R_{i1}, the gain with respect to input V_{i2} cannot exceed $(1 + R_f/R_{i1})$ which is obtained when $R_{i2} = 0$ (a direct connection).

A particularly useful case arises when $R_{i1} = R_{i2}$ and $R_f = R_g$. Equation (1.30) may be rewritten

$$V_o = V_{i2}\left(\frac{R_g}{R_{i2} + R_g}\right)\left(\frac{R_{i1} + R_f}{R_{i1}}\right) - V_{i1}\frac{R_f}{R_{i1}}. \tag{1.31}$$

For $R_{i1} = R_{i2} = R_i$ and $R_f = R_g$, this reduces to

$$V_o = (V_{i2} - V_{i1})\frac{R_f}{R_i}, \tag{1.32}$$

and the output is directly proportional to the difference between the two input voltages.

Suitable values of R_f and R_i can be determined using the computer program Input and Feedback Resistance Values (Inverting Amplifier) (Appendix A.1) with only a single input. If required, the program could be extended to cope with the unequal gain case of Equation (1.30). Attempts to exceed a gain of $(1 + R_f/R_{i1})$ for input V_{i2} will result in a negative value for R_{i2}, and a suitable check for this condition should be included in the program.

The output resistance of the differential amplifier will be very low, as in the configurations discussed previously. The definition of input resistance is more complicated since two distinct modes of operation are possible:

1. *Common mode operation*, where an input signal is applied to both V_{i1} and V_{i2} simultaneously. Neglecting current flow into the amplifier, the approximate effective input resistance is $(R_{i1} + R_f)$ in parallel with $(R_{i2} + R_g)$ or $(R_i + R_f)/2$ for the symmetrical case of Equation (1.32).

2. *Differential mode operation*, where the input signals change in opposite senses; since v_- is close to v_+, the approximate effective input resistance is $R_{i1} + R_{i2}$ or $2R_i$ for the symmetrical case.

1.5 Common Mode Rejection

Signals that appear simultaneously on both inputs of a differential amplifier are described as common mode. Those which appear as a difference between the two inputs are described as differential mode (sometimes "direct" or "series" mode).

In many practical applications, such as amplification of the output from a transducer that would be connected between V_{i1} and V_{i2} of Figure 1–12, the required signal is in differential mode. Any common mode signal that occurs on the two inputs is caused by the pick-up of interference on the leads connecting the transducer (and perhaps by the transducer itself). It is clearly important that the latter signal make the smallest possible contribution to the amplifier output signal; hence the importance of common mode rejection. Equation (1.32) implies that the common mode gain of the amplifier should be precisely 0 and the differential mode gain should be R_f/R_i as required.

Unfortunately, zero common mode gain assumes perfect matching of R_{i1} to R_{i2} and R_f to R_g. In practice this will not be so, and other spurious effects within the amplifier itself will combine to make the common mode gain small but *not* zero. Clearly, the smallness of this gain is a measure of the merit of the amplifier. This parameter is usually specified in inverse form as a *common mode rejection ratio* (CMRR) with respect to the differential mode gain and expressed in decibels; hence

$$\text{CMRR} = 20 \log_{10} \frac{(\text{Differential mode gain})}{(\text{Common mode gain})}. \tag{1.33}$$

The common mode rejection ratio is normally specified in manufacturers' data sheets for open loop operation. The quoted value will be further degraded in many applications by component mismatch. For this reason R_g, for example, in Figure 1–12 may consist of a fixed and a small variable resistance in series so that the latter can be adjusted for optimum common mode rejection.

In addition, although the common mode gain may be very small, the permissible magnitude of such a signal which may be applied to the amplifier inputs will have practical limits as specified by the manufacturer.

In general, therefore, the amplifier output will contain components due to both differential and common mode signals and can be written as

$$V_o = A_{dm}(V_{i2} - V_{i1}) + (A_{cm}/2)(V_{i2} + V_{i1}), \tag{1.34}$$

where the subscripts *dm* and *cm* refer to the differential and common mode gains, respectively. The factor 1/2 is normally included so that, for $V_{i2} = V_{i1}$, the common mode gain becomes A_{cm} (rather than $2A_{cm}$).

The values of A_{dm} and A_{cm} could be measured experimentally or derived for a particular circuit taking account of all relevant component tolerances. However, both these approaches yield gains with respect to V_{i1} and V_{i2}; that is, an expression of the form

$$V_o = A_{i2}V_{i2} - A_{i1}V_{i1} \tag{1.35}$$

(compare this with Equations [1.30] and [1.31]).

In order to determine the CMRR it is necessary to obtain A_{dm} and A_{cm} from A_{i1} and A_{i2} as follows. From Equation (1.34),

$$V_o = V_{i1}\,(A_{cm}/2 - A_{dm}) + V_{i2}\,(A_{cm}/2 + A_{dm}). \tag{1.36}$$

Compare this with Equation (1.35):

$$A_{i1} = -(A_{cm}/2 - A_{dm}) \tag{1.37}$$

and

$$A_{i2} = (A_{cm}/2 + A_{dm}). \tag{1.38}$$

Add (1.37) and (1.38):

$$A_{i1} + A_{i2} = 2A_{dm}. \tag{1.39}$$

Subtract (1.37) from (1.38):

$$A_{i2} - A_{i1} = A_{cm}. \tag{1.40}$$

Hence, from Equation (1.33), the common mode rejection ratio is given by

$$\text{CMRR} = 20 \log_{10} \frac{(A_{i2} + A_{i1})}{2\,(A_{i2} - A_{i1})}. \tag{1.41}$$

This value is computed by the computer program Common Mode Rejection Ratio (CMRR) Determination (Appendix A.3), which accepts gains as pure ratios and determines the CMRR in decibels.

This program clearly illustrates how closely the two gains must be matched in order to obtain really high common mode rejection ratios. For example, values of 99 and 100 for A_{i1} and A_{i2} (a difference of 1%) give a CMRR of 40 dB and 99.9 and 100 (a difference of 0.1%) give 60 dB; a difference of 0.01% is required to achieve 80 dB.

1.6 Instrumentation Amplifier

Although useful, the differential amplifier discussed in the previous section is unsuitable for applications involving high impedance sources because of its relatively low input resistance (determined essentially by the input resistors themselves).

The obvious remedy is to insert voltage followers (Figure 1–10) in each input path. Having incurred the expense of two extra amplifiers, it is tempting to operate these at a gain greater than unity, as shown in Figure 1–8. Unfortunately, in order to preserve common mode rejection, this approach would involve the matching of no less than four pairs of resistors. A clever interconnection of the buffer amplifiers avoids this requirement in the buffer stages. The resulting circuit is shown in Figure 1–13 and is known as an *instrumentation amplifier* because of its widespread use in measurement systems.

If the input current to the amplifiers can be neglected, the same current will flow through R_1, R_2 and R_3; let this current be I as shown in the figure. With voltages as indicated in the figure,

$$I = \frac{V_{o1} - V'_{i1}}{R_1} = \frac{V'_{i1} - V'_{i2}}{R_2} = \frac{V'_{i2} - V_{o2}}{R_3}. \tag{1.42}$$

From the first pair of terms,

$$\frac{V_{o1}}{R_1} = V'_{i1}\left(\frac{1}{R_2} + \frac{1}{R_1}\right) - \frac{V'_{i2}}{R_2}$$

$$\therefore V_{o1} = V'_{i1}\left(1 + \frac{R_1}{R_2}\right) - V'_{i2}\frac{R_1}{R_2}. \tag{1.43}$$

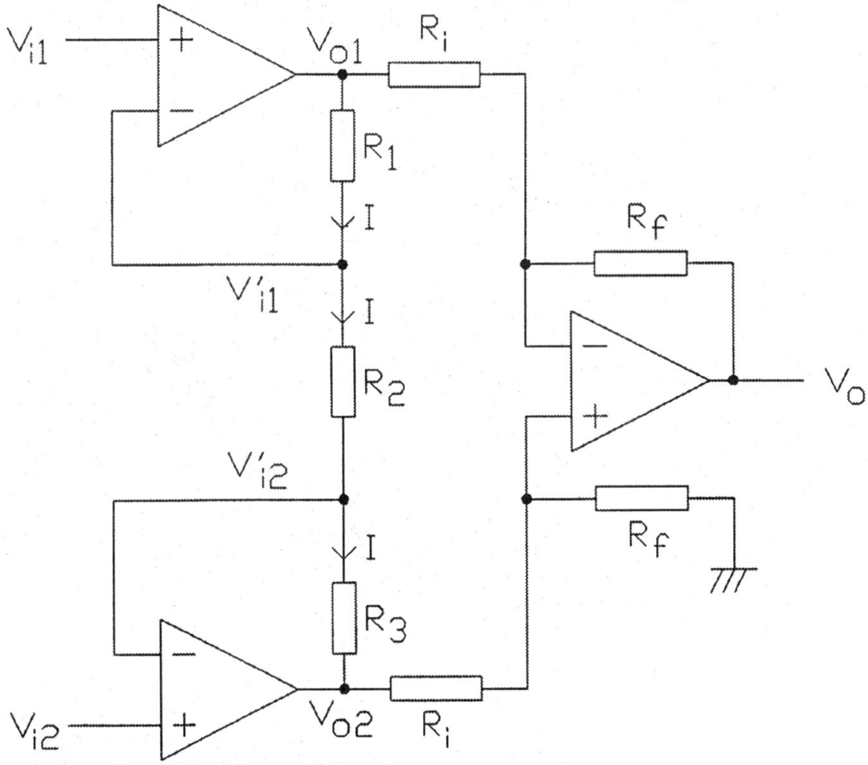

Fig. 1–13. Instrumentation amplifier.

Similarly, from the second pair of terms,

$$\frac{V_{o2}}{R_3} = V'_{i2}\left(\frac{1}{R_3} + \frac{1}{R_2}\right) - \frac{V'_{i1}}{R_2}$$

$$\therefore V_{o2} = V'_{i2}\left(1 + \frac{R_3}{R_2}\right) - V'_{i1}\frac{R_3}{R_2}. \tag{1.44}$$

In view of the high gain of the buffer amplifier, $V'_{i1} \approx V_{i1}$ and $V'_{i2} \approx V_{i2}$; hence, subtracting (1.44) from (1.43),

$$V_{o1} - V_{o2} = V_{i1}\left(1 + \frac{R_1}{R_2} + \frac{R_3}{R_2}\right) - V_{i2}\left(1 + \frac{R_1}{R_2} + \frac{R_3}{R_2}\right). \tag{1.45}$$

Notice that the coefficients of V_{i1} and V_{i2} are the same, and the expression reduces to

$$V_{o1} - V_{o2} = (V_{i1} - V_{i2})\left(1 + \frac{R_1 + R_3}{R_2}\right), \tag{1.46}$$

which ideally has zero common mode gain without any requirement for component matching. R_2 can conveniently be made variable in order to control the gain.

The overall gain including the output stage will be given by

$$V_o = (V_{i2} - V_{i1})\left(1 + \frac{R_1 + R_3}{R_2}\right)\frac{R_f}{R_i}. \tag{1.47}$$

It should be noted that the two resistors R_f in Figure 1–13 must be closely matched and ideally equal to each other; this applies also to the resistors R_i. As before, one of these components can be made adjustable in order to optimize common mode rejection.

Instrumentation amplifiers are available in the form of integrated circuit modules. These may contain resistances, the equivalent of R_2, which determine the gain and can be selected by means of links to give gains of 10, 100, or 1000; intermediate values can be obtained by using a resistor in place of the link.

If R_2 is removed, the input stages become unity gain buffers. However, if R_2 is short-circuited, Equation (1.46) suggests that the gain of the input stage should become infinite. This will not be the case, of course, since, for values of closed loop gain that approach the open loop value, the assumptions $V'_{i1} \cong V_{i1}$ and $V'_{i2} \cong V_{i2}$ are no longer valid. In practice, the circuit tends to oscillate when attempts are made to obtain very high gains.

1.7 Reference

Jung, W. J. 1986. *IC Op-Amp Cookbook.* Carmel, IN: Howard W. Sams.

CHAPTER 2

Frequency Response

2.1 Open Loop Behavior, Compensation

The circuits discussed in the previous chapter all depended on the assumption that the open loop gain *A* remains very large (ideally infinite) under all operating conditions. In practice, this cannot be true for all frequencies. For stable operation with the feedback configurations used, the high gain must be preserved for low frequencies, including dc (zero frequency).

However, for stable operation under all conditions, the gain must be made to fall or "roll off" at high frequencies. This will occur in any case due to stray capacitance, but additional capacitance is also used in order to define the frequency at which roll-off starts to occur. Roll-off is desirable not only to ensure stability but also to avoid amplification of signals outside the required range of frequencies, since this would merely increase the noise content.

This additional capacitance may be internal to the integrated circuit amplifier, in which case it would be realized as the barrier capacitance of a reverse-biased diode, or external, in which case it would be an orthodox capacitor. Both approaches are useful; the former is known as *internal compensation* and the latter as *external compensation*.

Internal compensation has the advantage that stability is guaranteed under all operating conditions and an external capacitor is not required. The disadvantage is that the available open loop bandwidth has been determined by the device manufacturer and cannot be readily changed by the user. The widely used 741 amplifier is of this type.

External compensation gives greater flexibility, but care is required since an unsuitable choice of compensating components can cause the amplifier to become unstable. The required values of compensating components are normally determined from manufacturers' data. The 748, for example, is essentially similar to the 741 except that it is designed for external compensation.

The simplest way of modeling this effect is by a single low pass filter, as shown in Figure 2–1. This first-order model is adequate for most applications, although a more accurate representation would include additional filter elements giving a higher-order transfer function.

The capacitance *C* of Figure 2–1 represents the total effect of all stray capacitances associated with the amplifier together with the compensating capacitance mentioned above. A_o represents the low frequency (dc) gain which, as usual, is very high. The second square is an ideal buffer amplifier whose gain may be assumed to be unity at all frequencies. The input to the unity gain buffer, and hence the output V_o, will be the input *v* amplified by A_o and modified by the potential divider action of *R* and *C*; that is,

$$V_o = A_o v \frac{1/j\omega C}{R + 1/j\omega C} = A_o v \frac{1}{1 + j\omega CR}, \tag{2.1}$$

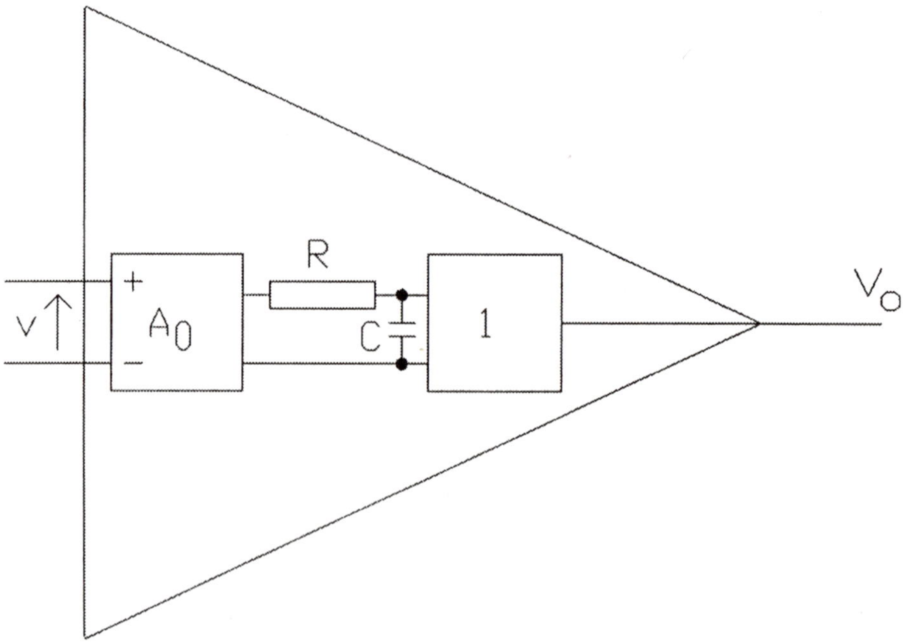

Figure 2–1. First-order model of amplifier behavior at high frequency.

where ω is the angular frequency (equal to $2\pi f$ where f is the frequency in hertz). For very low values of ω it can be seen that the gain tends to A_o, and for high values of ω it becomes small. The presence of a complex denominator means that there will also be a phase change with frequency. From Equation (2.1), the effective open loop gain is

$$A = \frac{V_o}{v} = A_o \frac{1}{1 + j\omega CR}. \tag{2.2}$$

This is often written

$$A = A_o \frac{1}{j(\omega/\omega_o)}, \tag{2.3}$$

where $\omega_o = 1/CR$.

It is customary to express the gain in decibels, hence

$$\text{Gain (dB)} = 20 \log_{10}|A| = 20 \log_{10} \frac{A_o}{\sqrt{1 + (\omega/\omega_o)^2}}. \tag{2.4}$$

It is useful to consider three cases:

1. $\omega \ll \omega_0$. The gain tends to $20 \log_{10} A_o$ which is the dc gain. So the graph of gain versus frequency will be a horizontal straight line under this condition.
2. $\omega \gg \omega_0$. This implies that ω/ω_o is much larger than unity, so Equation (2.4) becomes

$$\text{Gain (dB)} = 20 \log_{10} \frac{A_o}{(\omega/\omega_o)}, \tag{2.5}$$

which may be written

$$\text{Gain (dB)} = 20 \log_{10} A_o + 20 \log_{10} \omega_o - 20 \log \omega \tag{2.6}$$

The first two terms are constant with respect to ω and will therefore disappear if the ratio of the gains at two different frequencies is taken (since ratio implies the subtraction of logarithms). Consider two frequencies $\omega_1 = n\omega_o$ and

$\omega_2 = 10n\omega_o$, where n is large, so that $\omega_1 \gg \omega_o$ and $\omega_2 \gg \omega_o$. From Equation (2.6), the ratio of the gains at these two frequencies is given by

$$
\begin{aligned}
&\text{Gain (dB) at } \omega_2 - \text{Gain (dB) at } \omega_1\\
&= -20\log_{10}10n\omega_o - (-20\log_{10}n\omega_o)\\
&= -20\log_{10}10\\
&= -20 \text{ dB}
\end{aligned}
$$

The gain, therefore, decreases by 20 dB for each factor of ten increase in frequency. This is usually described as a slope, or roll-off, of 20 dB per decade. A similar argument, based on a doubling of the frequency, gives a slope of 6 dB per octave (where $\log_{10}2$ has been approximated to 0.3). So, at high frequencies, provided a logarithmic scale is used for both gain (by the use of dB) and frequency, a straight-line relationship is obtained with a (negative) slope as determined above. From Equation (2.6) this line will intersect the constant, low frequency, gain at $\omega = \omega_o = 2\pi f_o$.

3. $\omega = \omega_o$. Clearly for this condition neither of the two previous inequalities is valid. From Equation (2.4),

$$\text{Gain (dB)} = 20\log_{10}\frac{A_o}{\sqrt{2}} \tag{2.7}$$

$$= 20\log_{10}A_o - 20\log_{10}\sqrt{2}$$

$$= \text{dc gain} - 3 \text{ dB}. \tag{2.8}$$

Figure 2–2 is a computer-generated plot of Equation (2.4) for the 741 amplifier, which has a typical f_o of 5 Hz. Plots of this kind, usually known as "Bode plots," can be approximated at low frequencies by a horizontal straight line (corresponding to the dc gain and condition 1 above) and a straight line with a slope of –20 dB per decade at high frequencies (condition 2). These two straight lines intersect at f_o. Equation (2.8) shows that the gain at this point will actually be 3 dB less than the dc value as shown in Figure 2–2. This error is sufficiently small to be neglected in many applications; in any case, the straight lines can be drawn very simply once f_o (or ω_o) is known, and the required correction at the intersection point can be readily incorporated.

The frequency ω_o and its equivalent in Hertz ($f_o = \pi_o/2\omega$) are known variously as the *turnover frequency*, the *break frequency*, the *3 dB point* (from Equation [2.8]) and the *half power point* (since Equation (2.7) relates to voltage gain and $1/\sqrt{2}$ becomes squared when calculating power gain).

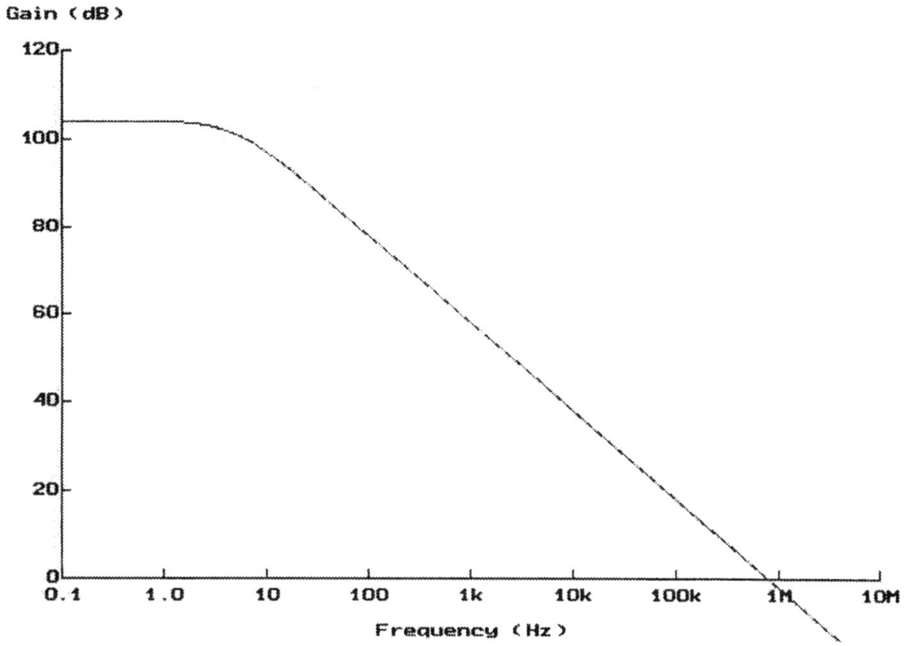

Figure 2–2. Bode plot for the open loop 741 amplifier (f_o = 5 Hz).

Figure 2–2 clearly illustrates the important concept of a constant "gain-bandwidth product" (GB). For all frequencies above f_o, the product of gain and frequency is constant; for example, voltage gains of 10^5 (100 dB) at 10 Hz, 10^3 (60 dB) at 1 kHz, and 1 (0 dB) at 1 MHz all lead to the same product:

$$10^5 \times 10 \text{ Hz} = 10^3 \times 1 \text{ kHz} = 1 \times 1 \text{ MHz} = 1 \text{ MHz} = \text{GB}.$$

2.2 Closed Loop Response, Rise Time

The ideas developed in the previous section relate to the amplifier itself. What happens when feedback is applied, as in Figure 1–3, in order to obtain a defined closed loop gain?

Clearly, at very low frequencies the assumption that A is very large will still be valid. But, as the frequency increases, this will become progressively less true until the required closed loop gain becomes equal to and then exceeds the available open loop gain. This is shown by the lower curve in Figure 2–3, which was obtained as follows from Equation (1–4) and enables allowance to be made for a finite open loop gain:

$$\frac{V_i + \frac{V_o}{A}}{R_i} + \frac{V_o + \frac{V_o}{A}}{R_f} = 0. \tag{1.4}$$

Separating V_o and V_i,

$$\frac{V_i}{R_i} + V_o\left(\frac{1}{A.R_i} + \frac{1}{R_f} + \frac{1}{A.R_f}\right) = 0. \tag{2.9}$$

Multiplying by R_f,

$$V_i\frac{R_f}{R_i} + V_o\left(\frac{1}{A}\frac{R_f}{R_i} + 1 + \frac{1}{A}\right) = 0 \tag{2.10}$$

$$\therefore \frac{V_o}{V_i} = -\frac{R_f}{R_i}\left(\frac{1}{1 + \frac{1}{A}\left(1 + \frac{R_f}{R_i}\right)}\right), \tag{2.11}$$

which reduces to $V_o/V_i = -R_f/R_i$ as usual for A large.

From Equation (2.4),

$$|A| = \frac{A_o}{\sqrt{1 + \left(\frac{\omega}{\omega_o}\right)^2}} = \frac{A_o}{\sqrt{1 + \left(\frac{f}{f_o}\right)^2}}. \tag{2.12}$$

Substituting into (2.11),

$$\frac{V_o}{V_i} = -\frac{R_f}{R_i}\left(\frac{1}{1 + \frac{1}{A_o}\sqrt{1 + \left(\frac{f}{f_o}\right)^2}\left(1 + \frac{R_f}{R_i}\right)}\right) \tag{2.13}$$

The closed loop gain shown in Figure 2–3 was computed using Equation (2.13) for $R_f/R_i = 100$ and expressing the gain in dB. At low frequencies, a closed loop gain of 40 dB is indeed obtained. At higher frequencies, the closed loop gain falls off until at the highest frequencies it coincides with the open loop characteristic.

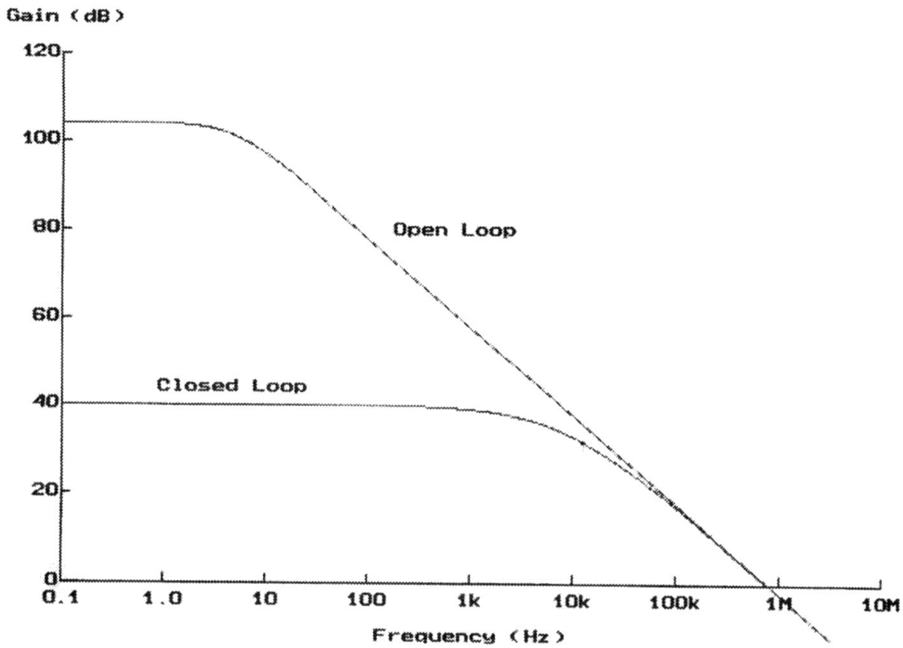

Figure 2–3. Closed loop gain of 100 (40 dB) superimposed on the open loop characteristic.

At approximately 10 kHz, in this case, the required closed loop gain of 40 dB would intersect the open loop characteristic. The actual closed loop gain is 3 dB less than its dc value at this frequency, as shown in Figure 2–3.

Consideration of Figure 2–3 shows that there is a trade-off between available closed loop gain and required bandwidth. If bandwidth is defined by a 3 dB drop in the closed loop value, it can be seen (from Figure 2–3) that

$$\text{(Closed loop gain)} \times \text{(Closed loop bandwidth)} = \text{GB}, \tag{2.14}$$

where GB is the gain-bandwidth product as defined in the previous section. For a given bandwidth, therefore, the maximum available closed loop gain follows immediately from the gain-bandwidth product of the amplifier to be used. For example, if a 741 amplifier (GB = 1 MHz) is to be used in an audio frequency application where a bandwidth of 20 kHz is needed, Equation (2.14) shows that the maximum available closed loop voltage gain is 50.

It should be noted that the bandwidth as defined by the 3 dB point (Equation [2.8]) actually represents an error of $1/\sqrt{2}$ (Equation [2.7]), or approximately 30% at the maximum frequency. Clearly, for precise applications such as instrumentation, much less bandwidth will be available. Acceptable gain errors for such applications are typically 1% or even 0.1%. The available bandwidth in these cases is readily determined from the 20 dB per decade roll-off of the amplifier.

For 1% error, the open loop gain must be at least 100 times the required closed loop gain. This corresponds to 40 dB, or two decades, so the available bandwidth is 1/100 of the 3 dB value. Similarly, 0.1% error reduces the bandwidth to 1/1000. Applying these values to the previous example of a 741 amplifier with a closed loop gain of 50, for a 1% gain accuracy the bandwidth becomes 200 Hz and for 0.1%, 20 Hz. This is a drastic reduction compared with the apparent implication of a 1 MHz gain-bandwidth product!

Although bandwidth is a convenient and widely used means of specifying an amplifier's behavior with respect to frequency, the concept of rise time is appropriate in circuits designed to handle square pulses. Since the precise start and finish of the rising edge are not well-defined, it is customary to specify the rise time as the time between 10% and 90% of the steady state output in response to an ideal step input, as shown in Figure 2–4.

The low pass filter effect of the amplifier implies that for narrow bandwidth amplifiers the rise time will be relatively long, and vice versa. The relationship can be calculated analytically but in practice the empirical relationship

$$\text{Rise time} = \frac{0.35}{\text{Gain-bandwidth product}} \tag{2.15}$$

(where rise time is defined as in Figure 2–4) is found to be particularly convenient. If the bandwidth is expressed in hertz, the rise time will be in seconds.

2.3 Large-Signal Operation, Slew Rate, and Full Power Bandwidth

The discussion of open and closed loop bandwidth and rise time in the previous section related to "small-signal operation." Although not specifically defined, this is generally assumed to refer to output signals of less than about 1 V peak to peak.

For large-signal operation, the output signal magnitude can approach the limits imposed by the power supply; for example, a swing of ± 12 V could be obtained with a conventional ± 15 V power supply. Under these conditions the factors that determine the effective bandwidth and rise time are somewhat different.

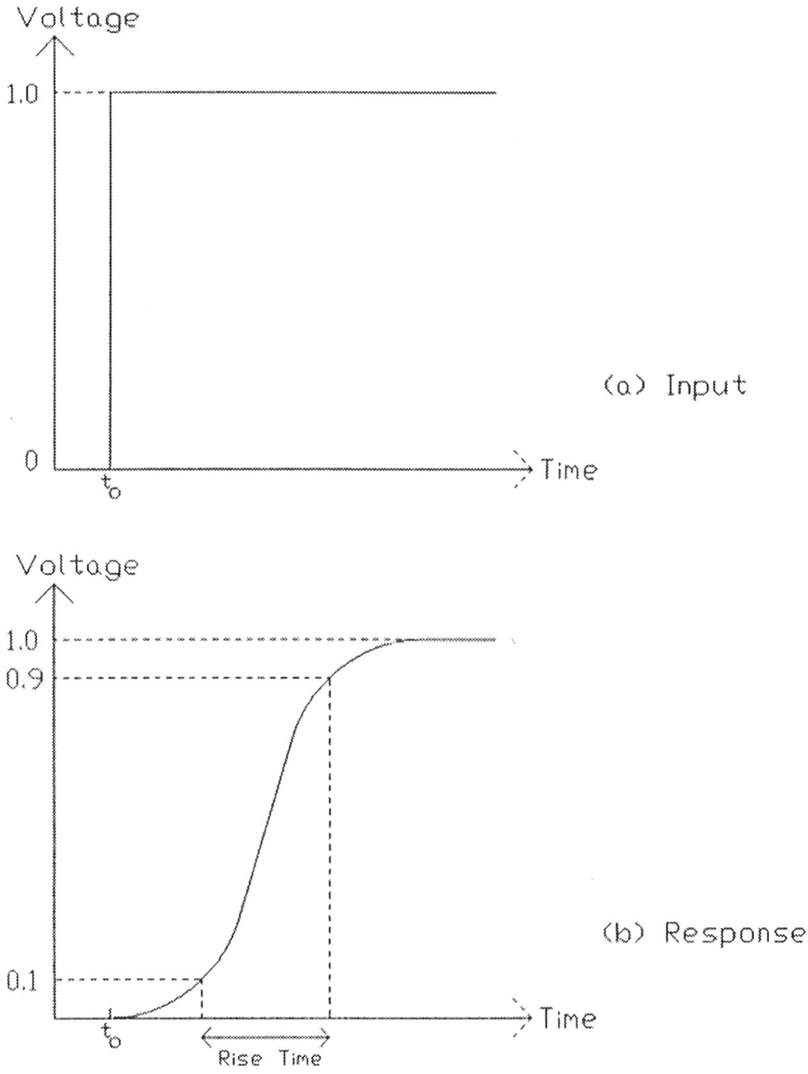

Figure 2–4. Definition of rise time.

The simple model of Figure 2–1 is still appropriate. However, high-frequency operation is now limited by the ability of amplifier A_o to provide sufficient current to charge capacitor C at the required rate. The corresponding rate of change of voltage at the output (dV_o/dt) when C is being charged at its maximum rate is called the *slew rate*, and an amplifier operating in this mode is said to be "slew-rate limited." For example, the familiar 741 amplifier has an effective value of C of 30 pF, and 15 μA (determined by the internal configuration of the amplifier) is available to charge this.

Now charge $Q = CV$, where V is the instantaneous voltage across any capacitor, by differentiating

$$\frac{dQ}{dt} = I = C\frac{dV}{dt}. \tag{2.16}$$

Inserting the values above, $dV/dt = 0.5$ V/μs, which is the maximum possible value and hence the slew rate.

The rise time for a 741 amplifier, for small-signal operation, may be obtained from Equation (2.15) as 0.35 μs. For slew-rate limited operation and an output step of height H, the effective rise time will be given by

$$\text{Rise time}_{\text{(effective)}} = \frac{0.8H}{\text{Slew rate}}, \tag{2.17}$$

where the 0.8 arises from the 10% to 90% definition of rise time (Figure 2–4). For the 741 amplifier and a 10 V step, the effective rise time becomes 16 μs, which is substantially longer than the small-signal value of 0.355 μs mentioned above.

Slew rate limiting will dominate when the step magnitude exceeds that value which makes the values given by Equations (2.15) and (2.17) equal. Clearly, therefore, the distinction between small- and large-signal operation depends on the parameters of the particular amplifier, and the 1 V value mentioned earlier is arbitrary. For the 741 amplifier, the values given by Equations (2.15) and (2.17) become equal for

$$\frac{0.35}{1.0} = \frac{0.8 \times H}{0.5} \text{V/μs} \tag{2.18}$$

or $H = 0.22$ V; the output will be slew-rate limited above this level.

Slew rate limiting also has important implications in the amplification of sinusoidal signals. In particular, at relatively high frequencies and amplitudes, the maximum rate of change of the sinusoid will be limited by the slew rate of the amplifier. Bandwidth defined in this way is called the *full power bandwidth* and can be less than the equivalent small-signal bandwidth.

Consider an output sine wave defined by

$$V_o = E_{pk} \sin \omega t, \tag{2.19}$$

where E_{pk} is the peak amplitude. The rate of change is given by

$$\frac{dV_o}{dt} = \omega E_{pk} \cos \omega t, \tag{2.20}$$

which has a maximum value of E_{pk} when cos ωt = 1 (where the original sinusoid crosses the zero level).

The highest possible output frequency without distortion due to slew rate limiting is therefore given by

$$\text{Slew rate} = \omega_{\max} E_{pk} = 2\pi f_{\max} E_{pk}. \tag{2.21}$$

Therefore

$$f_{\max} = \frac{\text{Slew rate}}{2\pi E_{pk}}. \tag{2.22}$$

For the 741 amplifier and a 20 V peak-to-peak (10 V peak) sine wave, the full power bandwidth is given by

$$f_{\max} = \frac{0.5 \times 10^6}{2\pi \times 10} = 7.96 \text{ kHz}. \tag{2.23}$$

Notice that this value depends on the output signal magnitude and not the gain (as in the case of small-signal operation). The dominant factor in determining the bandwidth depends on the required operating conditions; both small-signal and full-power bandwidths should always be checked for a proposed application.

Some manufacturers quote slew rate in their data sheets, some quote full-power bandwidth, and some quote both. The computer program Slew Rate to Bandwidth Conversion (Appendix A.4) provides convenient conversion between the two, taking account of the required signal amplitude.

CHAPTER 3

Offset Errors

3.1 Offset Voltage, Bias, and Difference Currents

Operation of the basic amplifier configurations, discussed in Chapter 1, was based essentially on three assumptions:

1. The open loop gain A is infinite.
2. The output voltage V_o is 0 when the net sum of the inputs is 0.
3. The current flowing into or out of the amplifier inputs is 0.

For a real amplifier, none of these will be precisely true. The effect of a finite value of A may readily be determined (see Equation [2.11]) and corrected by adjustment of the value of R_f and/or R_i; it will not be considered further.

The second effect above may be allowed for by means of a hypothetical *offset voltage* v_{os} that is normally referred to the amplifier inputs; this is shown together with the so-called *input bias currents* (the third effect) in Figure 3–1. A compensating resistor R_c has also been included and will be discussed later.

Summing currents at the inverting amplifier input,

$$\frac{V_i - v_-}{R_i} + \frac{V_o - v_-}{R_f} + i_{b-} = 0. \tag{3.1}$$

From Figure 3–1,

$$V_o = A\,(v_+ - v_- + v_{os}) \tag{3.2}$$

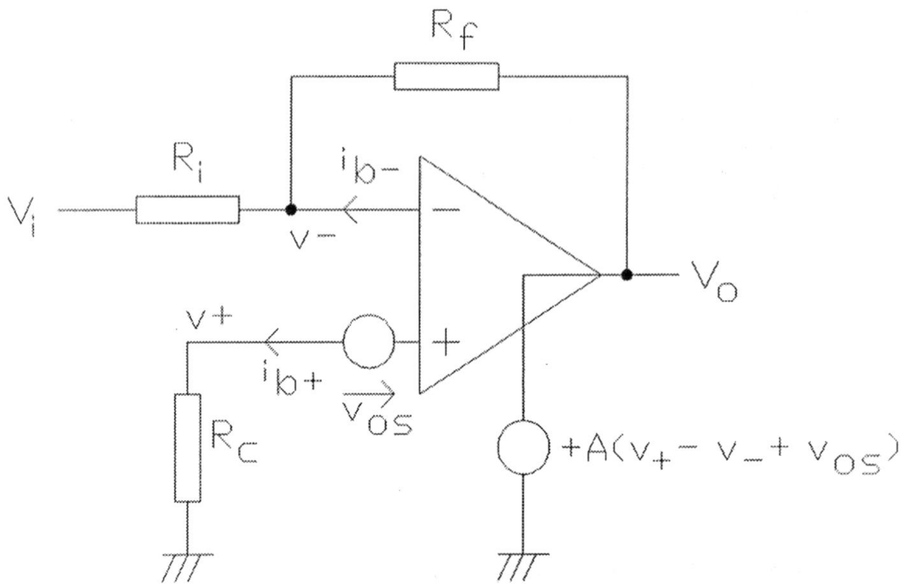

Figure 3–1. Operational amplifier circuit showing input offset voltage and bias currents.

$$\therefore \frac{V_o}{A} = v_+ - v_- + V_{os} \tag{3.3}$$

$$\therefore v_- = v_+ + v_{os} - \frac{V_o}{A}. \tag{3.4}$$

For A very large,

$$v_- \approx v_+ + v_{os}. \tag{3.5}$$

Substituting (3.5) into (3.1),

$$\frac{V_i - (v_+ + v_{os})}{R_i} + \frac{V_o - (v_+ + v_{os})}{R_f} + i_{b-} = 0. \tag{3.6}$$

Separating terms,

$$\frac{V_i}{R_i} - v_+\left(\frac{1}{R_i} + \frac{1}{R_f}\right) - v_{os}\left(\frac{1}{R_i} + \frac{1}{R_f}\right) + \frac{V_o}{R_f} + i_{b-} = 0. \tag{3.7}$$

Extracting V_o and multiplying by R_f,

$$V_o = -V_i\frac{R_f}{R_i} + v_{os}\left(1 + \frac{R_f}{R_i}\right) - R_f i_{b-} + v_+\left(1 + \frac{R_f}{R_i}\right). \tag{3.8}$$

The first term on the right-hand side of Equation (3.8) is the required one, the second is the output error due to the offset voltage, and the third is the output error due to input bias current. The last term is due to the presence of the compensating resistor R_c and is 0 if R_c is 0 (since v_+ then becomes 0). However, this term can be used to advantage as follows.

From Figure 3–1, $v_+ = R_c i_{b+}$; substituting in (3.8),

$$V_o = -V_i\frac{R_f}{R_i} + v_{os}\left(1 + \frac{R_f}{R_i}\right) - R_f i_{b-} + R_c i_{b+}\left(1 + \frac{R_f}{R_i}\right). \tag{3.9}$$

$$R_f = R_c\left(1 + \frac{R_f}{R_i}\right), \tag{3.10}$$

the coefficients of the i_{b+} and i_{b-} terms become equal and (3.9) reduces to:

$$V_o = -V_i\frac{R_f}{R_i} + v_{os}\left(1 + \frac{R_f}{R_i}\right) - R_f i_d, \tag{3.11}$$

where $i_d = (i_{b+} - i_{b-})$ is the (differential) *input offset current*. Since i_d is significantly smaller than i_{b+} and i_{b-} for most operational amplifiers, this reduction in offset error justifies the use of an additional resistor. From Equation (3.10), its value is given by

$$R_c = \frac{R_f R_i}{R_i + R_f},$$

which is the value of R_f and R_i in parallel. This means that the two amplifier inputs "see" the same source resistance and only the difference between the two bias currents (i_d) contributes to the offset at the output.

The computer program Offset Error Determination (Appendix A.5) calculates the total error due to the offset voltage and bias current, using Equation (3.8) for the uncompensated case (that is, the term involving v_+ is omitted). The magnitudes of the terms are added in order to obtain a "worst case" value; it is possible that the terms could be of opposite signs and, therefore, partially cancel. Subsequently, the required value of a compensating resistor is determined using Equation (3.11) and, for a specified input offset current, the total offset is recalculated for the compensated case.

The reduction in total offset due to the use of a compensating resistance can be demonstrated, and its usefulness may readily be assessed for a particular application. For example, with a feedback resistance R_f of 1 MΩ, an input resistance R_i of 10 kΩ (giving a gain of 100), an offset voltage v_{os} of 1 mV and an input bias current of 80 nA, the resultant total offset error at the amplifier output is 181 mV without a compensating resistance. This error reduces to 121 mV with the required compensating resistance of 9.9 kΩ for an input offset current of 20 nA. Although not highly effective, this technique is sometimes used and explains the presence of an otherwise puzzling resistance in the noninverting input of the amplifier.

It should be noted that, in the case of amplifiers that use field effect transistors in their input stages ("FET input types"), the improvement obtained by using a compensating resistor may be negligible in view of the very low input bias and difference currents.

It is also important to note that, since the current terms involve R_f but not R_i, a small value of R_f should be chosen for a particular required gain. This conflicts, however, with the requirements for high input resistance to avoid loading the previous stage, so a compromise must be made. The computer programs presented in this book (Appendix A.1 and A.5) should enable such a compromise to be readily achieved for a particular application; the T-network feedback configuration (Section 3.3) can also be useful.

3.2 Temperature and Other Effects

The offset voltages and bias currents, discussed in the previous section, have so far been assumed to be constant. In practice this is not the case. Both consist of a constant term, usually specified at 25°C, together with terms that specify the change with ambient temperature and supply voltage. Hence the offset voltage may be written

$$v_{os} = V_{os} + \frac{\delta v_{os}}{\delta T}\Delta T + \frac{\delta v_{os}}{\delta V_s}\Delta V_s, \tag{3.12}$$

and the bias current,

$$i_b = I_b + \frac{\delta i_b}{\delta T}\Delta T + \frac{\delta i_b}{\delta V_s}\Delta V_s, \tag{3.13}$$

where ΔT is the change in temperature and ΔV_s the change in supply voltage. Manufacturers normally specify the temperature parameters ($\delta v_{os}/\delta T$ and $\delta i_b/\delta T$) in microvolts (or nanoamps) per degrees Celsius and the supply voltage parameter ($\delta v_{os}/\delta V_s$) as a rejection ratio in microvolts (of offset) per volt (of supply change). The last term in Equation (3.13) is not normally specified and can usually be neglected.

The constant terms in Equations (3.12) and (3.13) do not present a problem. Provision can readily be made for their effects to be cancelled if required. This normally involves the connection of an external trimming potentiometer to the amplifier in a manner specified by the device manufacturer. This adjusts the current flows in the amplifier so that zero output can be obtained for zero input. Alternatively, a small, adjustable, compensating signal may be added using the configuration of Figure 1–4. A comprehensive treatment of offset adjustment is given by Jung (1986, 133–46).

Having cancelled the initial offset errors, it is still necessary to ensure that the chosen configuration does not exceed the required offset tolerances for all possible operating conditions. The computer program Worst Case Drift Performance (Appendix A.6) evaluates the worst case total offset, at the amplifier output, for the maximum anticipated temperature and power supply deviations. For simplicity, it is assumed that the nominal operating temperature is 25°C and the worst case temperature is *above* this value; if very low temperature operation is envisioned, a similar check should be incorporated for this case.

3.3 Use of T Network to Reduce Feedback Resistance

In order to retain a high value of input resistance when operating at high gain, an impractically high value of feedback resistance may be required. Also, the bias current contribution to the total offset is directly proportional to the feedback resistance (Equation [3.8]). It is, therefore, desirable to reduce the value of feedback resistance while retaining the required gain.

This reduction can be achieved by including an attenuator circuit, usually in the form of a T network, in the amplifier feedback path, as shown in Figure 3–2. The noninverting input is grounded as usual. It is convenient, but not essential, to make the two arms of the network (R_t) of equal value. Since the virtual ground point presents a very low resistance to ground, V'_o may be calculated from V_o by way of a potential divider network consisting of R_t together with R_t and R_s in parallel. Let the effective resistance of R_s and R_t in parallel be

$$R' = \frac{R_s R_t}{R_s + R_t}. \tag{3.14}$$

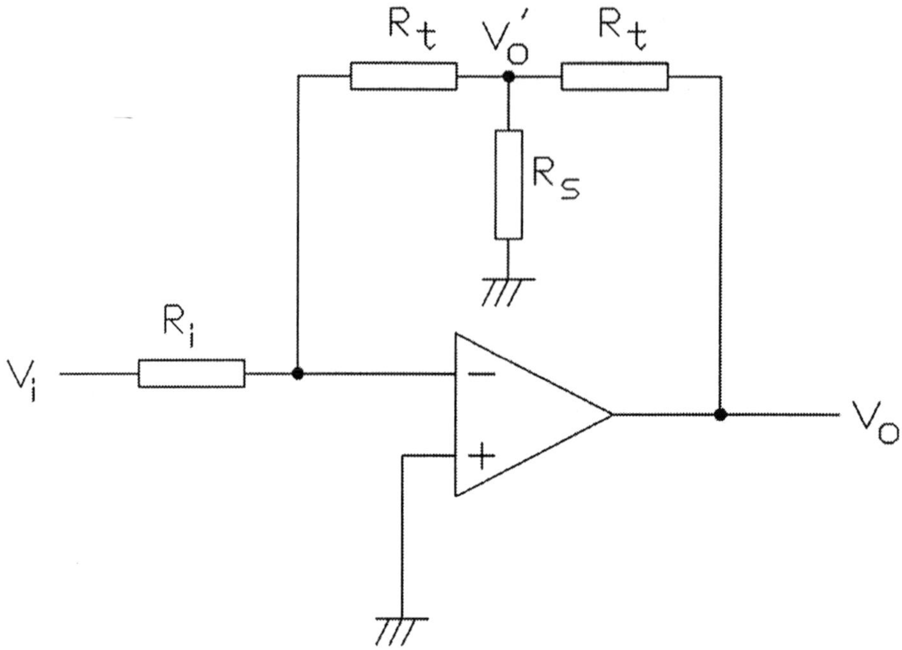

Figure 3–2. Operational amplifier with feedback attenuator.

Hence,

$$V'_o = V_o \frac{R'}{R_t + R'}. \tag{3.15}$$

Substituting (3.14) into (3.15),

$$V'_o = V_o \frac{\dfrac{R_s R_t}{R_s + R_t}}{R_t + \dfrac{R_s R_t}{R_s + R_t}}. \tag{3.16}$$

Multiplying by $(R_s + R_t)$,

$$V'_o = V_o \frac{R_s R_t}{R_s R_t + R_t (R_s + R_t)}. \tag{3.17}$$

Summing currents at the amplifier input in the usual way gives

$$\frac{V_i}{R_i} + \frac{V_o'}{R_t} = 0. \tag{3.18}$$

From (3.18),

$$V_o' = -\frac{V_i}{R_i} R_t. \tag{3.19}$$

Eliminating V_o' from Equations (3.17) and (3.19),

$$-\frac{V_i}{R_i} R_t = V_o \frac{R_s R_t}{R_s R_t + R_t (R_s + R_t)} \tag{3.20}$$

$$\therefore \frac{V_o}{V_i} = -\frac{R_s R_t + R_t (R_s + R_t)}{R_s R_i} \tag{3.21}$$

$$\therefore \frac{V_o}{V_i} = -\frac{1}{R_i}\left(\frac{R_t^2 + 2R_s R_t}{R_s}\right). \tag{3.22}$$

Comparing with Equation (1.6),

$$\frac{V_o}{V_i} = -\frac{R_f}{R_i}, \tag{1.6}$$

an effective feedback resistance $(R_f)_{eff}$ can be defined where

$$(R_f)_{eff} = \frac{R_t^2 + 2R_s R_t}{R_s}, \tag{3.23}$$

$(R_f)_{eff}$ is the value of feedback resistance that would be required in the conventional operational amplifier configuration (Figure 1–3). The required value of $(R_f)_{eff}$ is normally known, and R_s and R_t are to be determined. As there are two unknowns, one of them must be chosen arbitrarily. R_t is normally chosen, since choice of R_s would lead to a quadratic equation for R_t. Hence, from Equation (3.23),

$$R_s = \frac{R_t^2}{(R_f)_{eff} - 2R_t}. \tag{3.24}$$

The computer program T Network Feedback (Appendix A.7) evaluates Equation (3.24) for the required value of $(R_f)_{eff}$ (notice that megohms are used since large values can be expected) and an arbitrary choice of R_r. Obviously, the chosen value of R_t should be significantly smaller than $(R_f)_{eff}$; if it is not less than half this value, Equation (3.24) will fail since its denominator becomes negative. It may be necessary to try several values of R_t in order to obtain a convenient value for R_s.

The amplifier input bias current will now flow through the feedback network which, from Figure 3–2, can be seen to present a source resistance of R_t + (R_s and R_t in parallel). The value of R_s is usually small, so this approximates to R_r. The output error due to bias current flowing in this resistance is evaluated in the program and compared with that due to a single equivalent feedback resistance.

For example, an effective feedback resistance of 100 MΩ consisting of 47 Ω for R_t and the required value of 22.11 Ω for R_s gives an output offset error of 3.76 mV for a bias current of 80 nA. This compares with an error of 8 V with a single 100 MΩ feedback resistor. Although a carefully selected example, this clearly illustrates the improvement that can be achieved; the effective offset voltage at the output is reduced by a factor of approximately 2,000.

3.4 Blocking of DC Offset

Operation at very low frequencies, including dc, is an important and useful feature of operational amplifiers. However, their low cost and ease of use make them attractive for applications where very low frequency performance is not required (for example in audio amplifiers). In such cases, the effect of any offset may be eliminated by means of a capacitor, called a *blocking capacitor*, that could be connected directly in series with the amplifier output. This is shown, for the case of a two-stage amplifier, in Figure 3–3.

For $V_i = 0$, capacitor C will charge to the output offset voltage of amplifier 1 via R_{i2} and the virtual ground of amplifier 2. For subsequent, nonzero, values of V_i, the offset voltage stored on C will be effectively subtracted from the output of amplifier 1, thereby providing amplifier 2 with an input that has no dc component.

In multistage amplifiers without blocking capacitors, the offset of the early stages is amplified by the gain of all subsequent stages. This can clearly lead to saturation of the later stages when high gains are used. Blocking capacitors avoid this problem by restricting the effect of offset to individual stages. Two precautions must be observed in designing amplifiers of this kind:

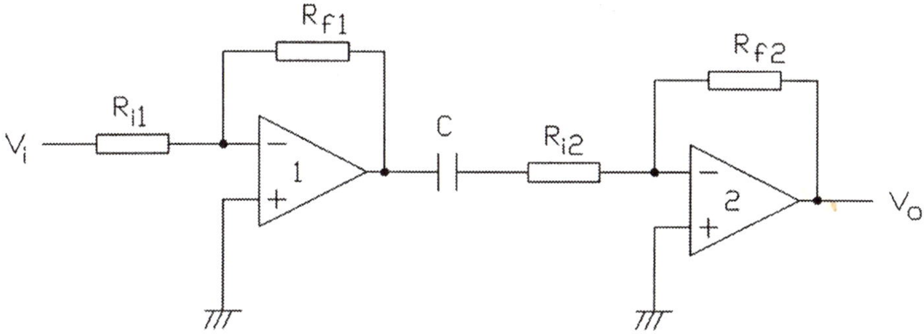

Figure 3–3. Two-stage operational amplifier circuit with blocking capacitor *C*.

1. The offset, even within a single stage, must not be so great that saturation can occur with the peak values of the anticipated alternating signal. The computer programs Offset Error Determination and Worst Case Drift Performance (Appendix A.5 and A.6) can be used to check for this.

2. The blocking capacitors must have sufficiently high values that the lowest required signal frequency is not attenuated excessively. The second stage of Figure 3–3 has effectively become the high pass filter of Figure 5–3, which will be discussed later.

More sophisticated operational amplifier circuits have been developed for ac amplification but are beyond the scope of this book. See, for example, Graeme et al. (1971, 222–25).

3.5 References

Graeme, J. G., G. E. Tobey, and L. P. Huelsman. 1971. *Operational Amplifiers—Design and Applications.* New York, McGraw-Hill.

Jung, W. J. 1986. *IC Op-Amp Cookbook.* Carmel, IN: Howard W. Sams.

CHAPTER 4

Waveform Generation

4.1 Preliminary Comments

Analog integrated circuits are particularly suitable for the generation of a wide range of waveforms. Several excellent devices are available specifically for the generation of timing pulses, triangular waves, sine waves, and suchlike; further details are available in the application notes of the appropriate manufacturers.

In this book, attention will be restricted to the use of operational amplifiers for the purpose of waveform generation. Before undertaking a detailed design for a specific application, it would be wise to check whether a ready-made device is available at a reasonable cost.

Although many different generation techniques are available, they may be broadly classified into those based on the generation of linear ramps (which may subsequently be shaped into sinusoidal or other forms by a suitable non-linear circuit) and those which generate sinusoids directly. The former are appropriate to applications where timing is the basic requirement and the latter to precision sine-wave generation, since harmonic distortion is difficult to remove from shaped triangular waves. These approaches will be discussed with examples in the following sections.

4.2 Ramp-Based Generators

Ramp-based generators depend on the use of one or more operational amplifiers arranged to integrate an input voltage with respect to time. A constant input voltage, therefore, produces a linear ramp output. The required configuration is

ration voltage (Figure 1–2) for the following reason: If V_i is initially made sufficiently positive, the voltage v_+ at the noninverting input will also be positive and, due to the high open loop gain of the amplifier, V_o will also be positive; in fact, V_o will be at its positive saturation voltage.

If V_i is made progressively more negative, the voltage v_+ will be reduced until eventually, as v_+ passes through zero, V_o will "flip over" to the negative saturation limit. Similarly, as V_i is made progressively more positive, V_o will "flip" back to its positive state. The condition for the voltage v_+ at the noninverting input to become zero can be determined as follows.

Let the current I flow through R_f and R_i as indicated in Figure 4–2 (current flow into the amplifier is neglected as usual). Hence,

$$I = \frac{V_o - V_i}{R_f + R_i} \tag{4.6}$$

and

$$v_+ = V_i + I.R_i$$

$$\therefore\ v_+ = V_i + \frac{V_o - V_i}{R_f + R_i} R_i. \tag{4.7}$$

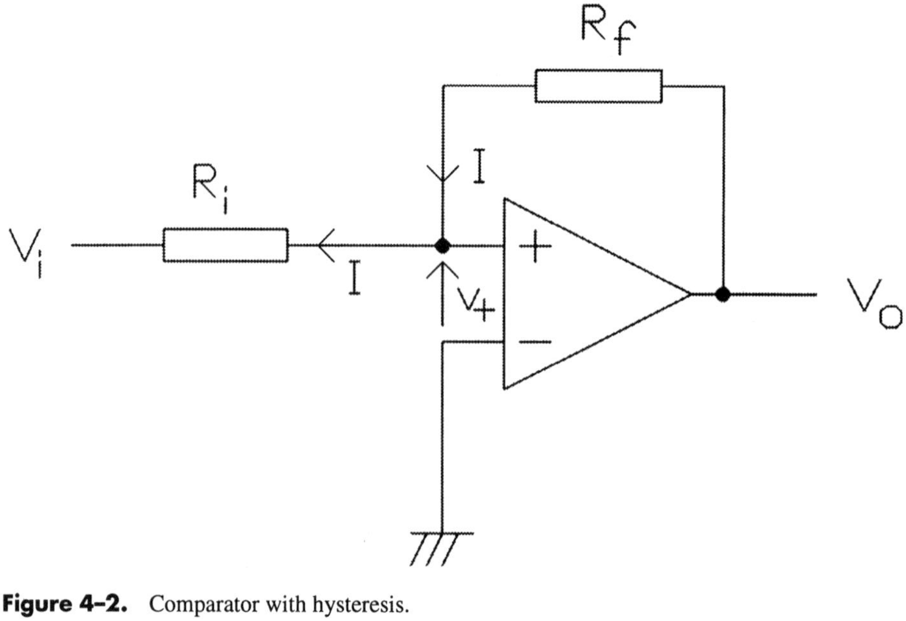

Figure 4–2. Comparator with hysteresis.

Hence, using $R_f + R_i$ as a common denominator,

$$v_+ = \frac{V_i(R_f + R_i) + (V_o - V_i)R_i}{R_f + R_i} \tag{4.8}$$

$$\therefore \; v_+ = \frac{V_iR_f + V_oR_i}{R_f + R_i}. \tag{4.9}$$

Hence $v_+ = 0$ when

$$V_i = -(V_o)_{sat}\frac{R_i}{R_f}, \tag{4.10}$$

where V_o has been replaced by $(V_o)_{sat}$, since the output will normally be in one of the two saturation states.

Normally, R_f will be made greater than R_i so that the condition for flipping from one saturation limit to the other, as specified by Equation (4.10), can be achieved with a value of V_i that is less than $(V_o)_{sat}$ and the comparator can be switched by a preceding amplifier with a similar output saturation voltage. If $R_f = nR_i$, the switching levels become $\pm(V_o)_{sat}/n$, as shown in Figure 4–3.

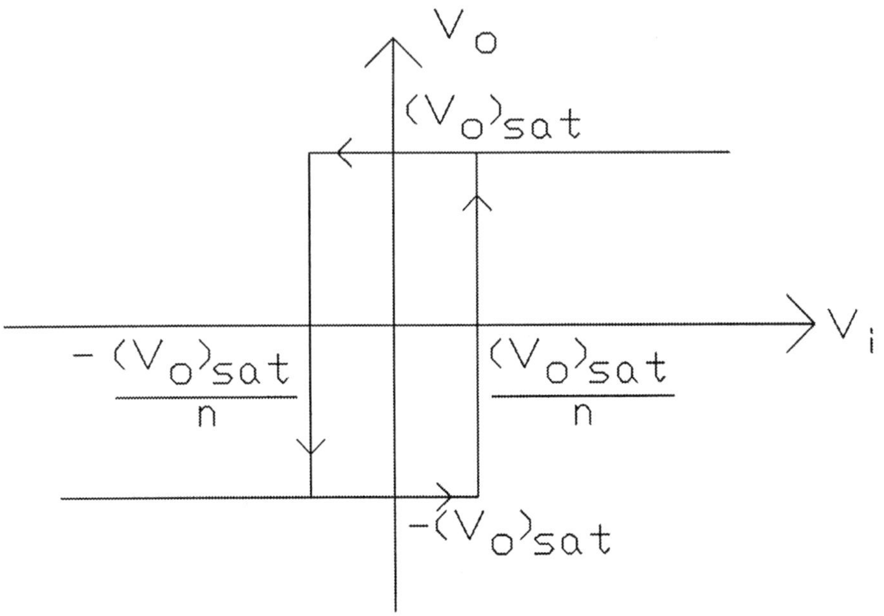

Figure 4–3. Characteristic of comparator with hysteresis.

The comparator with hysteresis can readily be used in conjunction with an operational integrator in order to provide a triangular and square-wave generator as shown in Figure 4–4(a). V_T will be a triangular wave with nearly equal positive and negative slopes and V_s will be a square wave that is positive during the falling ramp of V_T, and vice versa, as shown in Figure 4–4(b). The computer program Ramp Gen, Triangular Wave Oscillator Design, (Appendix A.8, page 116) enables suitable component values to be determined for particular operating conditions.

Matching of the positive and negative saturation voltages, and hence positive and negative slopes of V_T, can be improved by using Zener diodes and a

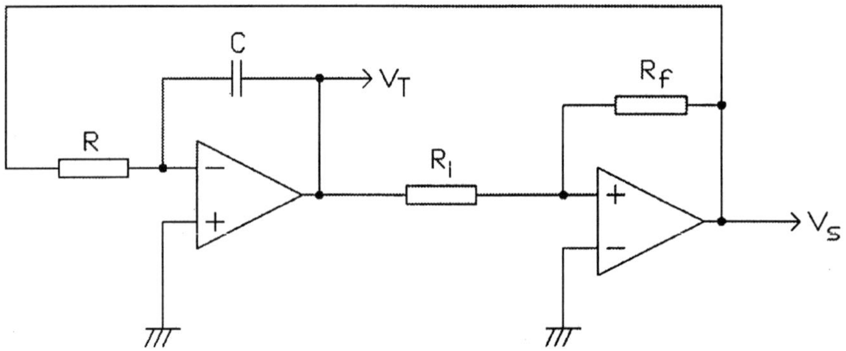

(a) Basic circuit

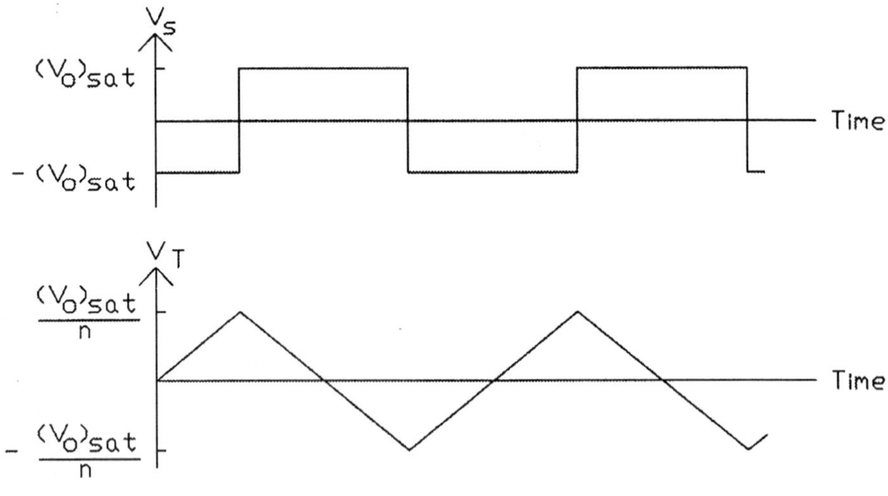

(b) Waveforms generated

Figure 4–4. Triangular waveform generator: (a) basic circuit; (b) waveforms generated.

current-limiting resistor to “clamp” the output of the comparator, as described by Jung (1986, 404) and indicated in Figure 4–5. The resistor R_{lim} is required to limit the amplifier output current when the Zener diodes are conducting. The limiting voltages become the operating voltages of the Zener diodes. These are stable to within a few percent, whereas amplifier saturation voltages can vary markedly. The overall dc level of the waveform may be adjusted by returning the inverting input of the comparator amplifier to an appropriate variable voltage.

Different gradients for the rising and falling ramps may be obtained by means of switches that select different integrator input resistances in the two cases; diodes can be used as suggested by Stout and Kaufman (1976, 27–4). Figure 4–6 shows the required circuit. When V_s is positive, the upper diode conducts and the input resistance of the integrator becomes $R_1 + R_3$, giving an integrator time constant of $(R_1 + R_3)C$. When V_s is negative, the lower diode conducts and the corresponding time constant becomes $(R_1 + R_2)C$.

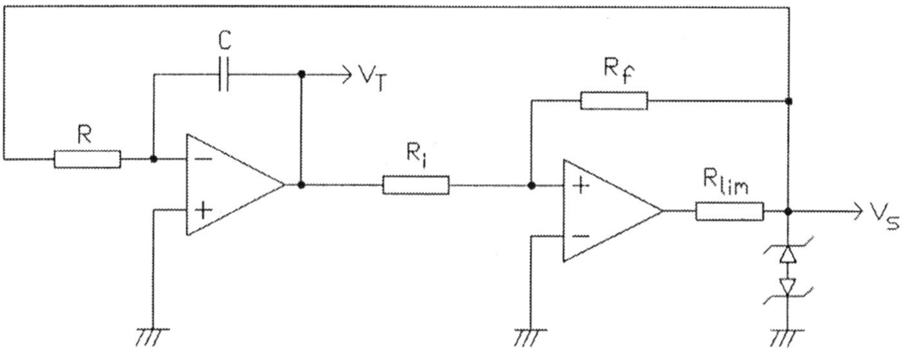

Figure 4–5. Improved triangular waveform generator with Zener diode voltage limiters.

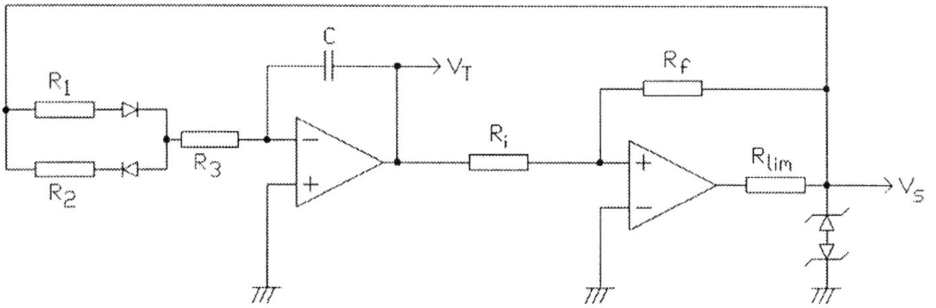

Figure 4–6. Triangular waveform generator with different positive and negative slopes.

In particular, the falling ramp may be made very fast, giving a good approximation to a sawtooth waveform, by using a high-speed switching device in order to discharge the integrating capacitor rapidly. A switching transistor may be used for this purpose, as described in detail by Jung (1986, 484–86).

4.3 Sine Wave Oscillators

Circuits in this category are characterized by positive feedback giving a closed loop gain very close to unity and by a frequency-sensitive network which ensures that the required conditions for oscillation are satisfied only at the required frequency. Many configurations have been used; only the *Wien-bridge*-based oscillator, which is widely used, will be discussed here.

The basic circuit is shown in Figure 4–7, where $R_1C_1R_2C_2$ constitute the Wien bridge and the amplifier provides a positive gain whose magnitude can

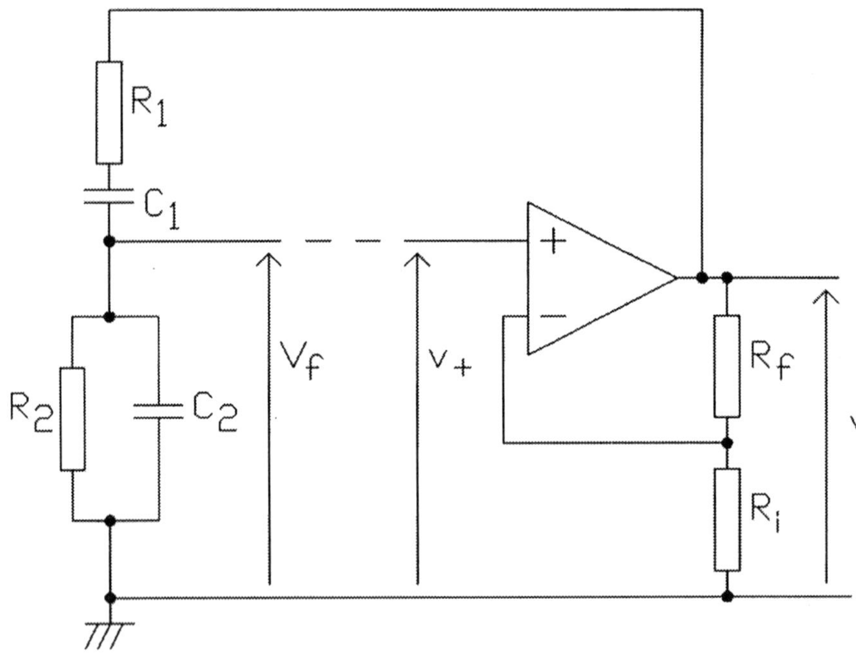

Figure 4–7. Basic Wien bridge oscillator.

be adjusted by means of R_f and R_i. The output V_f from the filter will be related to the amplifier output V_o by

$$V_f = V_o \frac{Z_2}{Z_1 + Z_2}, \tag{4.11}$$

where Z_1 is the impedance of R_1 and C_1 in series and Z_2 is the impedance of R_2 and C_2 in parallel; that is,

$$Z_1 = R_1 + \frac{1}{j\omega C_1}, \tag{4.12}$$

$$Z_2 = \frac{R_2 \dfrac{1}{j\omega C_2}}{R_2 + \dfrac{1}{j\omega C_2}} = \frac{R_2}{1 + j\omega C_2 R_2}. \tag{4.13}$$

Substituting (4.12) and (4.13) into (4.11) gives

$$V_f = V_o \frac{\dfrac{R_2}{1 + j\omega C_2 R_2}}{R_1 + \dfrac{1}{j\omega C_1} + \dfrac{R_2}{1 + j\omega C_2 R_2}}. \tag{4.14}$$

Multiplying by $(1 + j\omega C_2 R_2)$ gives

$$V_f = V_o \frac{R_2}{\left(R_1 + \dfrac{1}{j\omega C_1}\right)(1 + j\omega C_2 R_2) + R_2} \tag{4.15}$$

$$\therefore V_f = V_o \frac{R_2}{R_1 + R_2 + \dfrac{C_2 R_2}{C_1} + j\omega C_2 R_1 R_2 + \dfrac{1}{j\omega C_1}} \tag{4.16}$$

If $R_1 = R_2 = R$ and $C_1 = C_2 = C$, which is convenient but not essential,

$$V_f = V_o \frac{R}{3R + j\omega C R^2 + \dfrac{1}{j\omega C}}. \tag{4.17}$$

To produce oscillations of constant amplitude and frequency, the gain of the loop must be precisely one and there must be no net phase shift at the required frequency of oscillation. For there to be no net phase shift, the imaginary part of Equation (4.17) must be 0; hence,

$$j\omega CR^2 + \frac{1}{j\omega C} = 0. \tag{4.18}$$

Multiplying by j, and since $j^2 = -1$,

$$-\omega CR^2 + \frac{1}{\omega C} = 0 \tag{4.19}$$

$$\therefore\ \omega^2 C^2 R^2 = 1. \tag{4.20}$$

Hence, $\omega = 1/CR$, and

$$f = \frac{1}{2\pi RC}. \tag{4.21}$$

From Equation (1.18), the amplifier contributes a gain given by

$$V_o = v_+\left(1 + \frac{R_f}{R_i}\right). \tag{1.18}$$

From (1.18) and (4.17) (with the imaginary part set to 0), the loop gain from v_+ to V_f is given by

$$V_f = v_+\left(1 + \frac{R_f}{R_i}\right)\frac{R}{3R}. \tag{4.22}$$

When the loop is closed, $V_f = v_+$, and

$$1 = \left(1 + \frac{R_f}{R_i}\right)\frac{1}{3}; \tag{4.23}$$

therefore, $3R_i = R_i + R_f$ and $R_f = 2R_i$, which is the condition for a loop gain of one.

Unfortunately, this basic arrangement is not satisfactory since the required loop gain of *precisely* one cannot be maintained. Even a small decrease means that the oscillations die away, and any increase causes the oscillations to build up until saturation of the amplifier causes severe distortion. Many solutions to this problem have been suggested; all involve a nonlinear element of some kind, which ensures a gain slightly in excess of unity at low levels of oscillation and which falls below this value as the required amplitude of oscillation is exceeded. This ensures that oscillations build up when the circuit is first switched on and stabilize at the required amplitude. Unfortunately, any nonlinear network will cause some distortion of the generated sine wave, but, with careful design, this effect can be made small.

One possibility described by Jacob (1993, 297) is the use of the back-to-back diodes connected across R_f of Figure 4–7 in conjunction with an adjustable value of R_i, as shown in Figure 4–8. R_i is set to provide a loop gain slightly greater than unity. When the amplitude of oscillation causes the diode

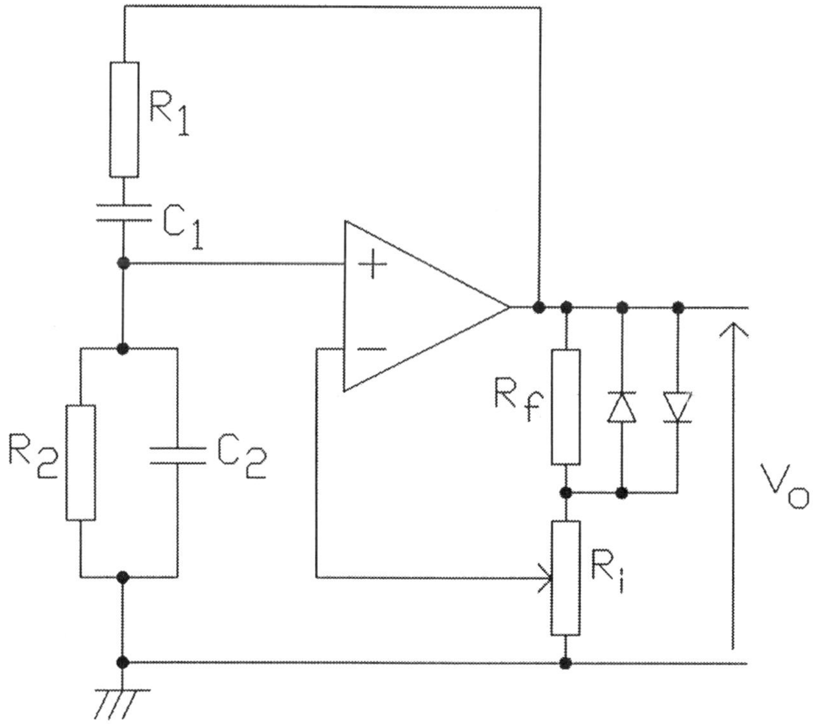

Figure 4–8. Wien bridge oscillator with simple amplitude stabilization.

forward voltage to be exceeded, the diodes begin to conduct, reducing the effective value of R_f and consequently the loop gain. Jacob (1993, 301) and Jung (1986, 451) describe an amplitude stabilizing circuit based on the use of a field effect transistor to reduce the loop gain at high signal levels. The high impedance input of this device minimizes the clipping effect on the generated waveform.

The computer program Wien Bridge Oscillator Design (Appendix A.9) enables the required component values for a specified frequency of oscillation to be determined using Equation (4.21).

4.4 References

Jacob, J. M. 1993. *Applications and Design with Analog Integrated Circuits.* Englewood Cliffs, NJ: Prentice-Hall International.

Jung, W. J. 1986. *IC Op-Amp Cookbook.* Carmel, IN: Howard W. Sams.

Stout, D. F., and M. Kaufman, eds. 1976. *Handbook of Operational Amplifier Circuit Design.* New York: McGraw-Hill.

CHAPTER 5

Introduction to Active Filters

5.1 Preliminary Comments

Filters are networks whose gain (or attenuation) and associated phase shift vary with frequency. This frequency dependency can be exploited in order to separate wanted and unwanted components of a signal on the basis of frequency. In general, filters consist of linear elements such as resistors, capacitors, and inductors, although useful filters can be constructed using any two of these. Inductors are widely used at radio frequencies where the required values are a convenient size. At lower frequencies, resistance-capacitance filters tend to predominate in view of the physical size and cost of the components required to produce sufficiently high values of inductance.

Filters may be broadly classified as passive or active. The former use resistive and reactive components alone and cannot, therefore, produce power gain; active filters use these components in conjunction with some form of amplifying circuit and can, therefore, produce power gain. Operational amplifiers are particularly suitable for this application.

Passive filters have the advantages of physical simplicity, operation at almost any required frequency, and lack of power supply requirements. A disadvantage is that source and load impedances must be taken into account when calculating the characteristics of the filter, since these effectively form part of the network. If these impedances change, the filter characteristics may change markedly. Also, component interactions mean that the filter must be analyzed as a whole and this often leads to involved mathematics. Active filters, on the other hand, can use buffer amplifiers in order to isolate each section of the filter from source and load impedance effects (see Section 1.3, page 10). Buffers between stages of filters mean that transfer functions can be cascaded by simple multiplication; this simplification is shown in Figure 5–1.

Provided the voltage followers are ideal,

$$V_o = V_s \frac{1}{(1 + j\omega C_1 R_1)\ (1 + j\omega C_2 R_2)}, \tag{5.1}$$

where V_o is the output signal, V_s is the input, or source, signal, and the terms in brackets represent the transfer functions of the individual filter networks (see Equation [2.1]). Notice that R_s and R_L, the source and load resistances, do not appear in this expression.

A further advantage of active filters is that the high input impedance of amplifiers means that long time constants can be obtained by using high resistance values, without the need for very large capacitance values. Of course, active filters cannot operate at frequencies where the constituent amplifiers no longer conform sufficiently accurately to the assumptions of ideal operation. For these reasons, active filters tend to be used at relatively low frequencies, with passive circuits being used at higher frequencies.

Filters may also be classified by their "order." The order is determined essentially by the number of reactive components in the filter, which in turn determines the order of the equation that relates the output of the filter to its input (the *transfer function*). The example in Figure 5–1 is, therefore, a second-order filter obtained, in this case, by cascading two first-order filters.

It has already been shown (Section 2.1) that the gain of a first-order low pass filter falls off at high frequencies by 20 dB per decade, regardless of its detailed configuration. In general, at high frequencies, the gain of an nth-order filter will fall off at $20n$ dB per decade. In a similar manner, the low-frequency gain of a high pass filter falls off by $20n$ dB per decade at low frequencies.

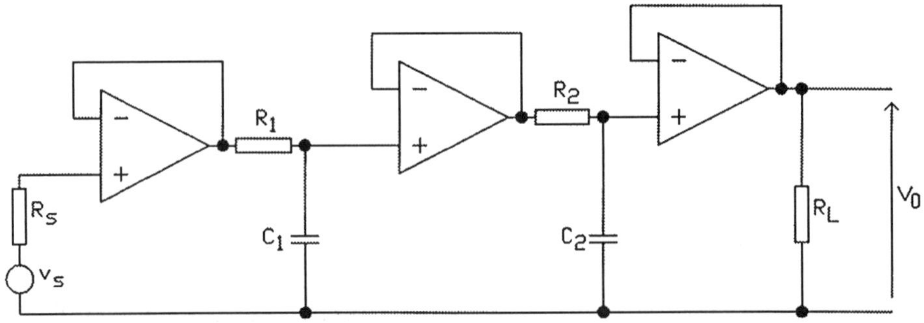

Figure 5–1. Cascaded filter networks.

The minimum order of filter required to separate two components of different frequency by a specified amount on this basis can, therefore, readily be determined. This is achieved by the computer program Filter Order Determination (Appendix A.10). However, it should be noted that high Q filters can exceed this performance for a particular range of frequencies (see Section 5.3.3 and Figure 5–11).

High-order filters can be produced in various ways. However, in many cases, it is convenient to use cascaded second-order filters (as discussed in Section 5.3) with a single first-order filter (Section 5.2) if the overall required order is odd. It is important to notice that cascaded first-order filters (Figure 5–1) do not provide the flexibility of specification of a true second-order arrangement and therefore cannot be used as the basic "building blocks."

5.2 First-Order Active Filters

First-order filters contain only one reactive component. A simple active, low pass filter may be produced using one of the RC sections shown in Figure 5–1 together with input and output buffer amplifiers. Its gain/frequency response will be that of the first-order model discussed in Section 2.1 and of the form shown in Figure 2–2. The break frequency is $\omega_o = (1/RC)$ and the low-frequency gain tends to unity.

A more convenient arrangement is shown in Figure 5–2; this configuration allows low-frequency gains other than unity to be obtained. The analysis which led to Equation (1.6) may easily be generalized to cover impedances by writing Z in place of R. In this case, Z_f becomes R_f in parallel with C_f, and R_i is unchanged. Hence,

$$Z_f = \frac{R_f \dfrac{1}{j\omega C_f}}{R_f + \dfrac{1}{j\omega C_f}} = \frac{R_f}{1 + j\omega C_f R_f}. \tag{5.2}$$

Therefore,

$$\frac{V_o}{V_i} = -\frac{Z_f}{Z_i} = -\frac{R_f}{R_i}\frac{1}{1 + j\omega C_f R_f}. \tag{5.3}$$

This shows that the amplifier has a low-frequency gain ($\omega \to 0$) defined, as usual, by $-(R_f/R_i)$ in conjunction with the response of a low pass filter of time constant $C_f R_f$ that does not depend on R_i. This implies that several input signals

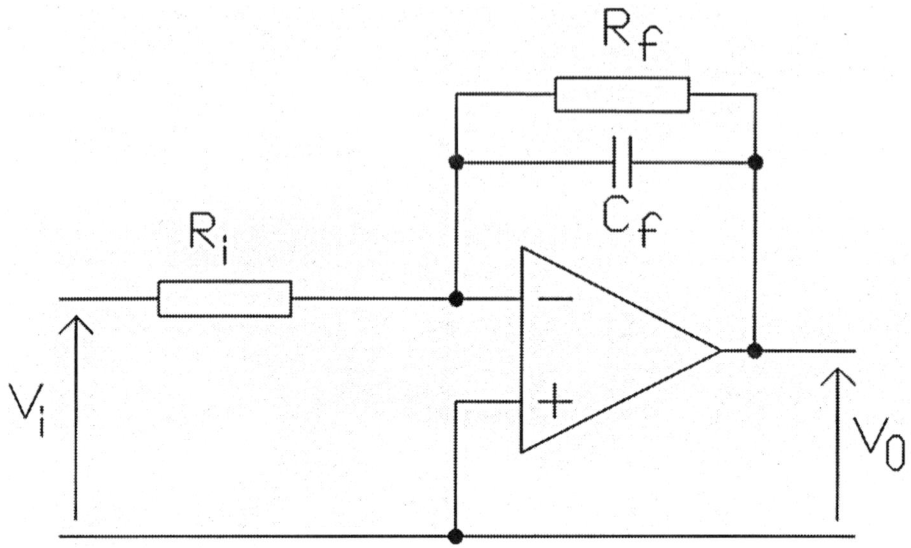

Figure 5–2. Active first-order low pass filter.

could be summed, with different weights determined by different values of R_i if required, and the break frequency would be the same for each input.

Rationalizing Equation (5.3) gives

$$\frac{V_o}{V_i} = -\frac{R_f}{R_i}\left(\frac{1 - j\omega C_f R_f}{1 - (j\omega C_f R_f)^2}\right)$$
$$= -\frac{R_f}{R_i}\left(\frac{1 - j\omega C_f R_f}{1 + \omega^2 C_f^2 R_f^2}\right). \tag{5.4}$$

The absolute value or modulus of a complex number $a + jb$ is given by $\sqrt{(a^2 + b^2)}$;hence,

$$\left|\frac{V_o}{V_i}\right| = \frac{R_f}{R_i}\left(\frac{1}{1 + \omega^2 C_f^2 R_f^2}\right)\sqrt{1 + \omega^2 C_f^2 R_f^2}$$
$$= \frac{R_f}{R_i}\frac{1}{\sqrt{1 + \omega^2 C_f^2 R_f^2}}. \tag{5.5}$$

Similarly, the phase angle ϕ represented by $a + jb$ is given by $\tan \phi = b/a$; hence,

$$\begin{aligned} \tan\phi &= -\omega R_f C_f \\ \therefore \phi &= -\arctan \omega R_f C_f. \end{aligned} \tag{5.6}$$

Equation (5.6) shows that the phase angle is 0 for very low frequencies, tends to −90° (i.e., lagging) for very high frequencies, and is −45° at the break frequency given by $\omega = 1/R_f C_f$, so that $\omega R_f C_f = 1$.

The configuration shown in Figure 5–3 enables a high pass characteristic to be obtained. As before,

$$\frac{V_o}{V_i} = -\frac{Z_f}{Z_i} = -\frac{R_f}{R_i + 1/j\omega C_i} = -\frac{j\omega C_i R_f}{1 + j\omega C_i R_i}. \tag{5.7}$$

From this it can be seen that, at very low frequencies, the gain tends to 0 and, at very high frequencies, when the 1 in the denominator becomes insignificant, to $-(R_f/R_i)$.

The interpretation of Equation (5.7) is simplified by the fact that Bode plots (Section 2.1), being logarithmic, may be added when transfer functions

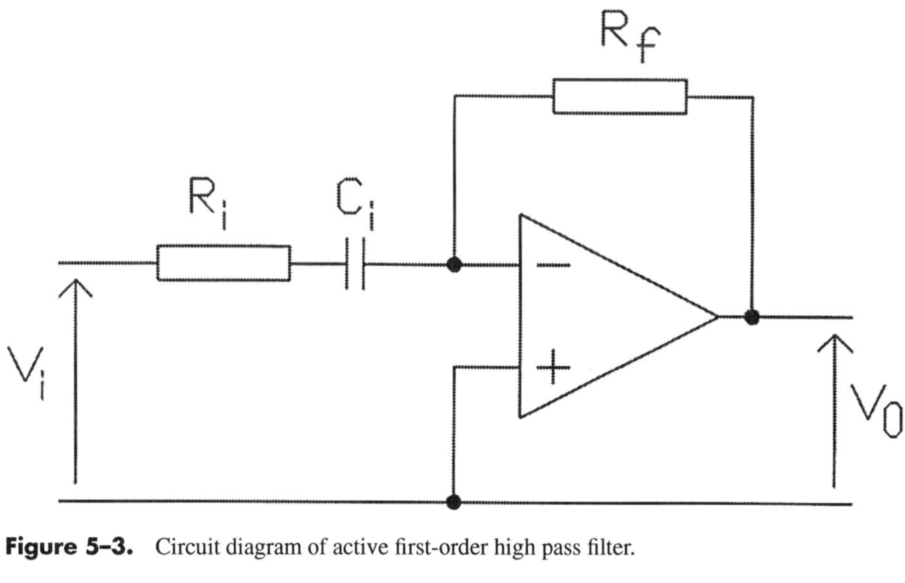

Figure 5–3. Circuit diagram of active first-order high pass filter.

are cascaded. The numerator represents a gain/frequency characteristic with a constant rise of 20 dB per decade; it will cross the 0 dB axis when $\omega C_i R_f = 1$ (at f_1, say, where $f_1 = 1/2\pi C_i R_f$) as shown by curve (1) in Figure 5–4. The denominator represents a low pass function with break frequency given by f_o where $f_o = \omega_o/2\pi$ and $\omega_o = 1/C_i R_i$ as shown by curve (2). These two curves may be added to give the solid line, shown in the figure, which is clearly a high pass characteristic. At high frequencies, the positive and negative slopes of 20 dB per decade cancel and give a constant gain that depends on the ratio of R_f and R_i and can be equal to, greater than, or less than unity.

The phase characteristics may also be added; the numerator will contribute 90° lead at all frequencies. When combined with the low pass response, the overall effect is 90° lead at low frequencies, 45° lead at the break frequency ($f_o = 1/2\pi C_i R_i$), and zero phase shift at high frequencies.

The input arrangement of Figure 5–3 may be combined with the feedback arrangement of Figure 5–2 to produce a *band pass filter* which will have a gain that approaches $-(R_f/R_i)$ over the required range of frequencies. The gain will fall off at 20 dB per decade at both low and high frequencies. Notice that, for

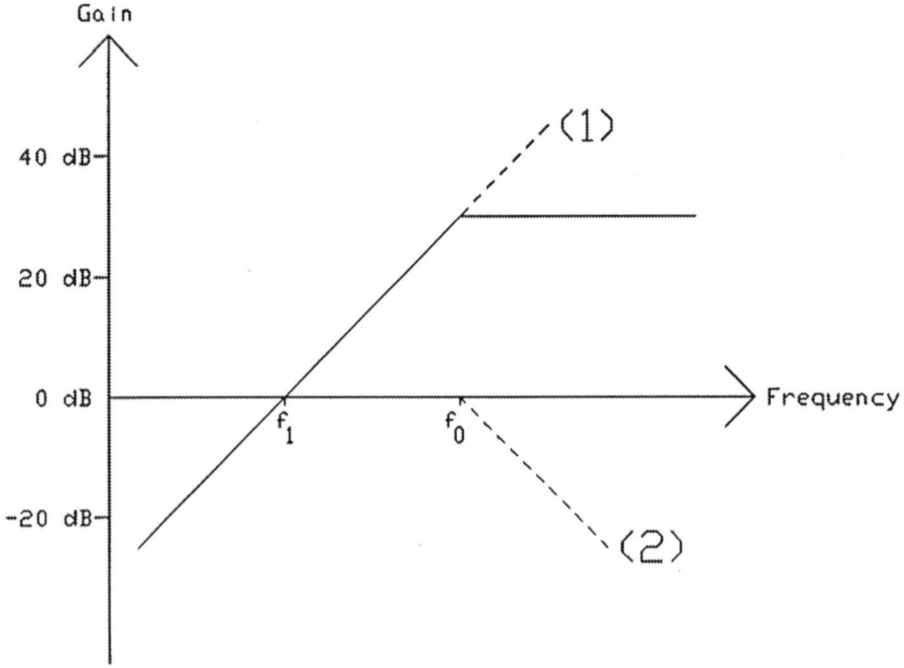

Figure 5–4. Bode plot for active first-order high pass filter.

narrow bandwidths, if the two break frequencies are sufficiently close to each other, the maximum gain will be less than $(-R_f/R_i)$(see also Section 5.3.3).

5.3 Second-Order Active Filters

As explained in Section 5.1, second-order active filters, based on a single operational amplifier, are useful both in their own right and as building blocks for higher-order filters. Although many configurations are possible, that shown in general form in Figure 5–5 is widely used. The components selected for each Z determine the form of the filter response (low, high, or band pass), as will be shown later. In most cases (the band pass filter is the only exception), each consists of a single resistor or capacitor, as appropriate. The resistances R_f and R_i determine the maximum available gain and may be omitted if unity is acceptable (in this case, the amplifier output would be connected to the inverting input either directly or via a resistance equal to that "seen" by the noninverting input in order to reduce bias current effects, as discussed in Section 3.1).

Analysis of this circuit does not involve any new concepts but is, nevertheless, slightly tortuous; the following approach is based on that used by Jacob (1993, 380–82). The relevant voltages and currents are identified in

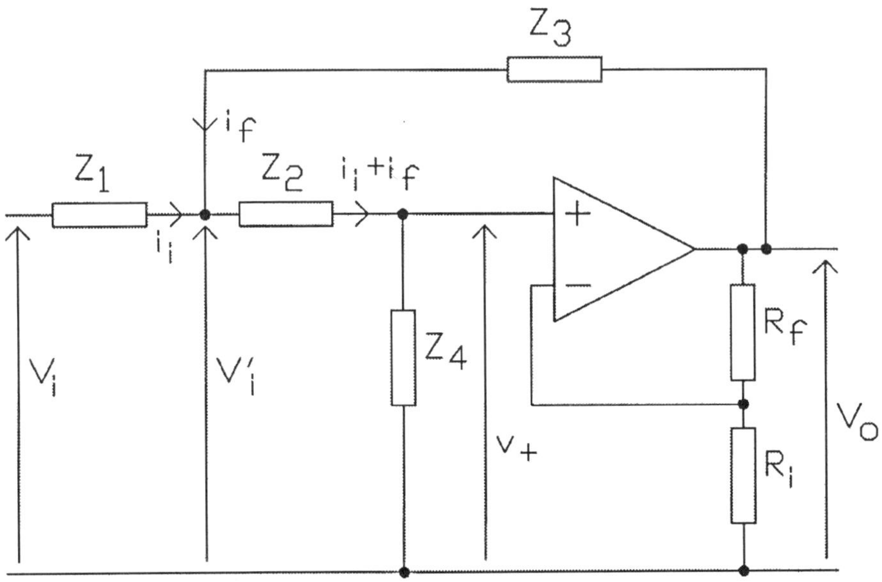

Figure 5–5. General second-order active filter.

Figure 5–5; if current flow into the amplifier is neglected, as usual, the current flowing through both Z_2 and Z_4 will, by Kirchhoff's current law, be the sum of the input current i_i and the feedback current i_f, as indicated. Applying Ohm's law to the impedance Z_1,

$$i_i = \frac{V_i - V_i'}{Z_1} = \frac{V_i}{Z_1} - \frac{V_i'}{Z_1}. \tag{5.7}$$

By the potential divider action of Z_2 and Z_4,

$$v_+ = V_i' \frac{Z_4}{Z_2 + Z_4} \tag{5.8}$$

$$\therefore V_i' = v_+ \frac{Z_2 + Z_4}{Z_4}. \tag{5.9}$$

Substituting V'_i as given by Equation (5.9) into Equation (5.7),

$$i_i = \frac{V_i}{Z_1} - v_+ \frac{Z_2 + Z_4}{Z_1 Z_4}. \tag{5.10}$$

Applying Ohm's law to impedance Z_4,

$$i_i + i_f = \frac{v_+}{Z_4} \tag{5.11}$$

$$\therefore i_i = \frac{v_+}{Z_4} - i_f. \tag{5.12}$$

Substituting this value for i_i into Equation (5.10),

$$\frac{v_+}{Z_4} - i_f = \frac{V_i}{Z_1} - v_+ \frac{Z_2 + Z_4}{Z_1 Z_4}. \tag{5.13}$$

Applying Ohm's law to impedance Z_3,

$$i_f = \frac{V_o - V_i'}{Z_3}. \tag{5.14}$$

Substituting this value for i_f into Equation (5.13),

$$\frac{v_+}{Z_4} - \frac{V_o - V_i'}{Z_3} = \frac{V_i}{Z_1} - v_+\frac{Z_2 + Z_4}{Z_1 Z_4} \tag{5.15}$$

$$\therefore \frac{v_+}{Z_4} - \frac{V_o}{Z_3} + \frac{V_i'}{Z_3} = \frac{V_i}{Z_1} - v_+\frac{Z_2 + Z_4}{Z_1 Z_4}. \tag{5.16}$$

Substituting for V'_i using Equation (5.9),

$$\frac{v_+}{Z_4} - \frac{V_o}{Z_3} + v_+\frac{Z_2 + Z_4}{Z_3 Z_4} = \frac{V_i}{Z_1} - v_+\frac{Z_2 + Z_4}{Z_1 Z_4}. \tag{5.17}$$

Now $V_o = A_o v_+$ where $A_o = 1 + R_f/R_i$ is the gain of the noninverting feedback amplifier as given by Equation (1.18):

$$\therefore v_+ = \frac{V_o}{A_0}. \tag{5.18}$$

Substituting this value of v_+ into Equation (5.17),

$$\frac{V_o}{A_0 Z_4} - \frac{V_o}{Z_3} + \frac{V_o}{A_0}\frac{Z_2 + Z_4}{Z_3 Z_4} = \frac{V_i}{Z_1} - \frac{V_o}{A_0}\frac{Z_2 + Z_4}{Z_1 Z_4}. \tag{5.19}$$

Separating terms in V_o and V_i,

$$V_o\left(\frac{1}{A_0 Z_4} - \frac{1}{Z_3} + \frac{Z_2 + Z_4}{A_0 Z_3 Z_4} + \frac{Z_2 + Z_4}{A_0 Z_1 Z_4}\right) = \frac{V_i}{Z_1}. \tag{5.20}$$

Multiplying through by Z_1,

$$V_o\left(\frac{Z_1}{A_0 Z_4} - \frac{Z_1}{Z_3} + \frac{Z_1 (Z_2 + Z_4)}{A_0 Z_3 Z_4} + \frac{Z_2 + Z_4}{A_0 Z_4}\right) = V_i. \tag{5.21}$$

Taking $A_0 Z_3 Z_4$ as a common denominator,

$$V_o\left(\frac{Z_1 Z_3 - A_0 Z_1 Z_4 + Z_1 (Z_2 + Z_4) + Z_3 (Z_2 + Z_4)}{A_0 Z_3 Z_4}\right) = V_i \tag{5.22}$$

$$\therefore \frac{V_o}{V_i} = \frac{A_0 Z_3 Z_4}{Z_1 Z_2 + Z_2 Z_3 + Z_3 Z_4 + Z_1 Z_3 + Z_1 Z_4 (1 - A_o)} \tag{5.23}$$

where $A_o = 1 + R_f/R_i$, as explained above.

5.3.1 Low Pass Filter

To produce a low pass filter, Z_1 and Z_2 are made resistive (R_1 and R_2 respectively), and Z_3 and Z_4 are made capacitive (C_1 and C_2 respectively). Equation (5.23) therefore becomes

$$\frac{V_o}{V_i}(s) = \frac{A_0 \frac{1}{sC_1}\frac{1}{sC_2}}{R_1 R_2 + R_2 \frac{1}{sC_1} + \frac{1}{sC_1}\frac{1}{sC_2} + R_1 \frac{1}{sC_1} + R_1 (1 - A_0) \frac{1}{sC_2}}, \tag{5.24}$$

where s is the Laplace operator (see, for example, Dorf 1992, 45–52), which may, for the purposes of this book, be regarded as equivalent to $j\omega$. Multiplying by $s^2/(R_1 R_2)$,

$$\frac{V_o}{V_i}(s) = \frac{\frac{A_0}{R_1 R_2 C_1 C_2}}{s^2 + \frac{s}{R_1 C_1} + \frac{1}{C_1 C_2 R_1 R_2} + \frac{s}{R_2 C_1} + (1 - A_0)\frac{s}{R_2 C_2}}. \tag{5.25}$$

Combining terms in accordance with their power of s,

$$\frac{V_o}{V_i}(s) = \frac{\frac{A_0}{R_1 R_2 C_1 C_2}}{s^2 + s\left(\frac{1}{R_1 C_1} + \frac{1}{R_2 C_1} + \frac{1 - A_0}{R_2 C_2}\right) + \frac{1}{R_1 R_2 C_1 C_2}}, \tag{5.26}$$

which is of the general form

$$\frac{V_o}{V_i}(s) = A_o \frac{\omega_o^2}{s^2 + 2\zeta\omega_o s + \omega_o^2}, \tag{5.27}$$

where ω_o is called the *critical frequency* and ζ is the damping factor whose significance is explained below. This equation has been studied very widely in connection with linear dynamic systems (particularly in relation to feedback control; see, for example, Dorf 1992, 169).

For the second-order active low pass filter specified by Equation (5.26), comparing the terms in the denominator with a coefficient s^o,

$$\omega_o^2 = \frac{1}{R_1 R_2 C_1 C_2}$$
$$\therefore \omega_o = \frac{1}{\sqrt{R_1 C_1 R_2 C_2}}. \tag{5.28}$$

Similarly, comparing denominator terms with a coefficient of s,

$$2\zeta\omega_o = \frac{1}{R_1 C_1} + \frac{1}{R_2 C_1} + \frac{1 - A_0}{R_2 C_2}. \tag{5.29}$$

Substituting for ω_0 as given by (5.28),

$$2\zeta \frac{1}{\sqrt{R_1 C_1 R_2 C_2}} = \frac{1}{R_1 C_1} + \frac{1}{R_2 C_1} + \frac{1 - A_0}{R_2 C_2}. \tag{5.30}$$

Multiplying both sides by $R_1 C_1 R_2 C_2$,

$$2\zeta\sqrt{R_1 C_1 R_2 C_2} = R_2 C_2 + R_1 C_2 + (1 - A_0)\, R_1 C_1 \tag{5.31}$$

$$\therefore 2\zeta = \frac{R_2 C_2 + R_1 C_2 + (1 - A_0)\, R_1 C_1}{\sqrt{R_1 C_1 R_2 C_2}}. \tag{5.32}$$

Filter theory 2ζ, as defined in this case by Equation (5.32), is often written as α (or some other symbol) and referred to as a *damping coefficient*. For all values of α, the second-order low pass active filter has the following characteristics:

a. low-frequency gain of A_0,
b. high-frequency gain which falls off at 40 dB per decade (twice the rate of a first-order filter),
c. zero phase shift at very low frequencies,
d. 180° (lagging) phase shift at high frequencies, and
e. 90° (lagging) phase shift at the natural frequency (ω_0).

The only, but important, effect of α is on the nature of the gain and phase characteristics in the vicinity of ω_0 (where $\omega/\omega_o = f/f_o = 1$). Those values of α that have particular properties have been specifically named as follows:

1. *Bessel filter.* In some applications it is desirable that all frequency components of the signal should be delayed by an equal amount. It can be shown (see, for example, Brown and Glazier 1974, 80) that this occurs when the variation of phase with frequency is linear. The second-order Bessel filter ($\alpha = 1.73$) closely meets this requirement. However, its amplitude response starts to fall off relatively early in the pass band and its transient response is heavily damped.

2. *Butterworth filter.* This gives the flattest amplitude response before roll-off starts to occur and is widely used for this reason. The transient response, however, can show a slight overshoot. For a second-order Butterworth filter, the value of α is $\sqrt{2} = 1.414$. Since $\alpha = 2\zeta$, this corresponds to $\zeta = 1/\sqrt{2}$, which is often regarded as providing the fastest response for a linear, second-order control system (see Dorf 1992, 179). Gain/frequency and phase/frequency plots for second-order Bessel and Butterworth filters are shown in Figure 5–6.

3. *Chebyshev filters.* These have relatively low values of α and are characterized by a peak in the gain/frequency response followed by a rapid *initial* fall-off (the final value will be 40 dB/decade, as usual); the change of phase is also more abrupt. The rate of initial fall-off increases as the height of the peak increases; values above 3 dB are not normally used since lightly damped filters of this kind have a tendency to become unstable.

The required values of α are obtained from the standard expression for the peak gain (Dorf 1992, 314):

$$|G|_{\max} = \frac{1}{2\zeta\sqrt{1-\zeta^2}} \quad \text{for } \zeta < 0.707. \tag{5.33}$$

Hence, since $\alpha = 2\zeta$,

$$|G|_{\max} = \frac{1}{\alpha\sqrt{1-\dfrac{\alpha^2}{4}}} \quad \text{for } \alpha < 1.414. \tag{5.34}$$

Since this involves solution of a fourth-order equation, the most frequently required values are summarized for convenience in Table 5–1.

The peak in the response occurs at a frequency ω_p that is *not* the same as ω_0. These two frequencies are related by the following standard expression (see, for example, Dorf 1992, 314):

$$\omega_p = \omega_o\sqrt{1-2\zeta^2}. \tag{5.35}$$

Since $2\zeta = \alpha$, $\zeta = \alpha/2$ and

$$2\zeta^2 = 2\frac{\alpha^2}{4} = \frac{\alpha^2}{2}. \tag{5.36}$$

Therefore, from Equation (5.35),

$$\omega_p = \omega_0\sqrt{1-\frac{\alpha^2}{2}}, \tag{5.37}$$

and

$$\omega_0 = \frac{\omega_p}{\sqrt{1-\frac{\alpha^2}{2}}}. \tag{5.38}$$

These features are illustrated in the gain/frequency and phase/frequency plots of Figure 5–7.

Equation (5.26) shows that, in order to define these two parameters, no less than six component values must be determined (since A_0 is determined by R_f and R_i). Following the pioneering work of Sallen and Key (1955), two approaches have been developed that greatly reduce the required design effort.

Table 5–1. Damping coefficients for a range of Chebyshev filters.

Height of peak relative to dc (dB)	*0.1*	*0.25*	*0.5*	*1*	*2*	*3*
Damping coefficient (α)	1.303	1.236	1.158	1.046	0.8862	0.7666

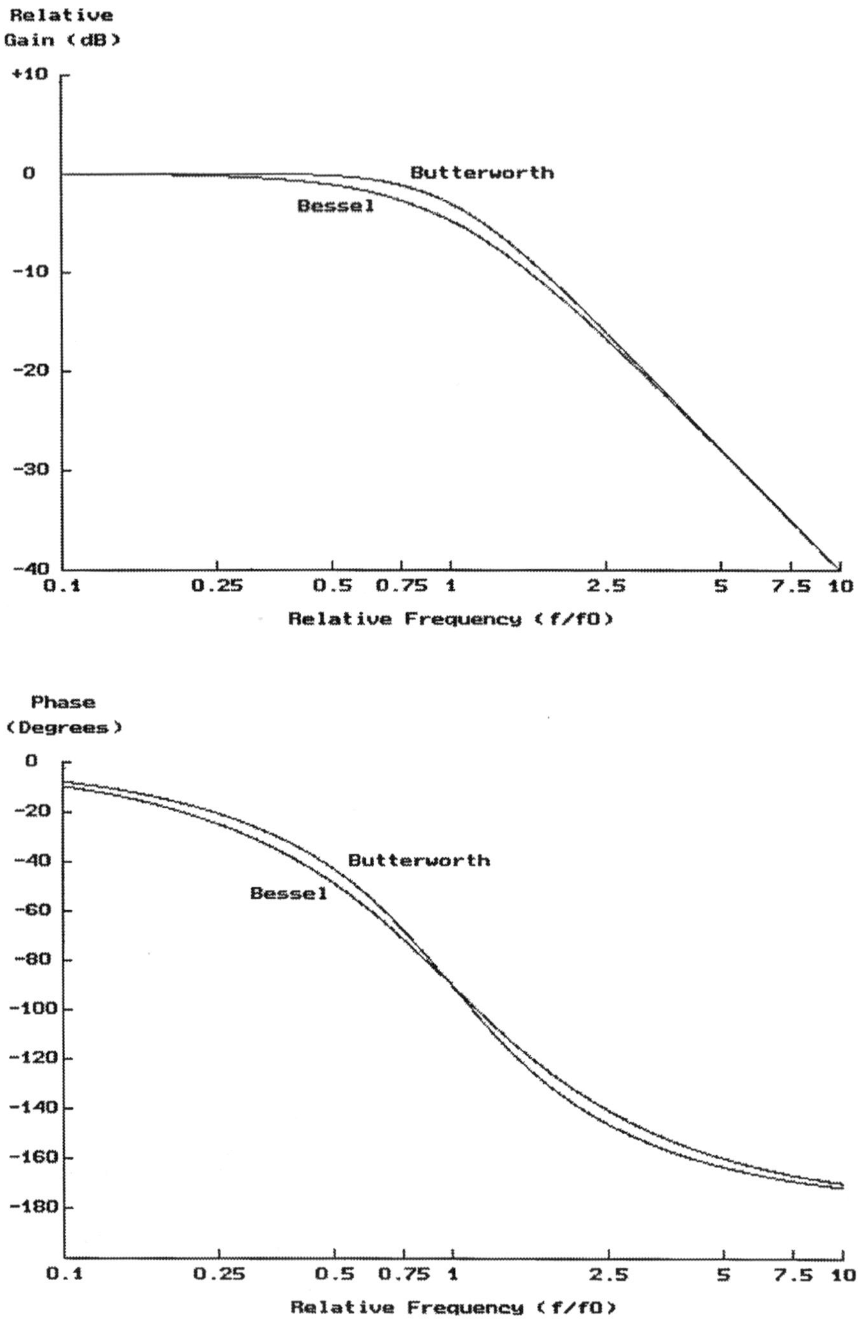

Figure 5–6. Second-order Bessel ($\alpha = 1.732$) and Butterworth ($\alpha = 1.414$) low pass filters: (a) gain (dB) versus frequency (logarithmic scale); (b) phase (degrees) versus frequency (logarithmic scale).

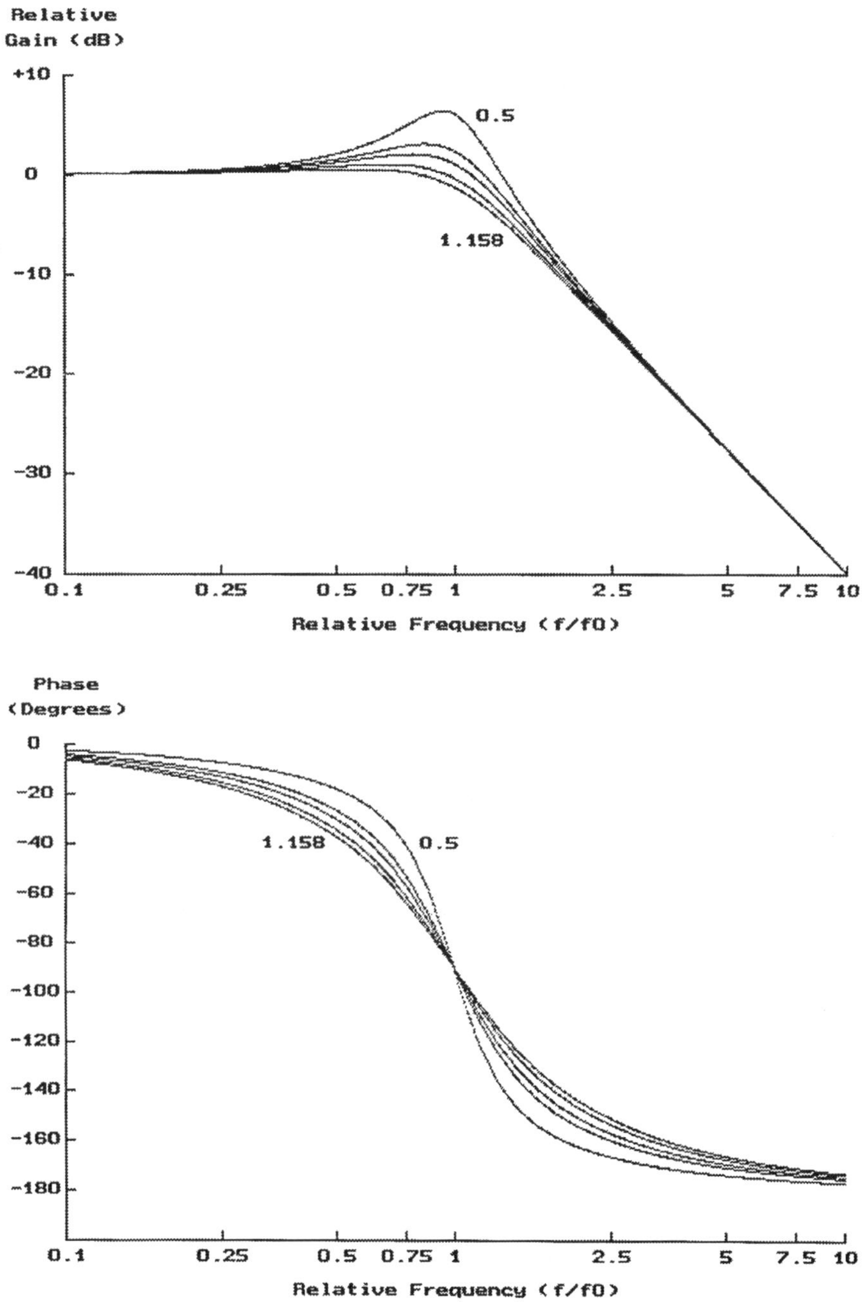

Figure 5–7. Second-order Chebyshev low pass filters for $\alpha = 0.5, 0.766, 0.866, 1.046$, and 1.158: (a) gain (dB) versus frequency (logarithmic scale); (b) phase (degrees) versus frequency (logarithmic scale).

(1) Unity Gain Filters

In this case, A_0 is made equal to 1 (the amplifier operates as a simple voltage follower), and the two remaining resistors are made equal to each other ($R_1 = R_2 = R$).

From Equations (5.28) and (5.32) and since $\alpha = 2\zeta$,

$$\omega_o = \frac{1}{R\sqrt{C_1C_2}} \tag{5.39}$$

$$\alpha = 2\sqrt{C_2/C_1}. \tag{5.40}$$

The computer program Unity Gain Sallen-Key Filter Design (Appendix A.11) makes use of these equations in order to determine suitable component values.

(2) Equal Component Filter

Although Equations (5.39) and (5.40) were easy to apply, difficulties can arise in view of the *ratio* of two capacitances, which determines α. A convenient value for C_1 may lead to an awkward one for C_2, and for low damping coefficients the required ratio can become large.

An alternative is to assign the same values to the two resistors and also to the two capacitors; that is, $R_1 = R_2 = R$ and $C_1 = C_2 = C$. The value of A_0 is no longer restricted to unity, of course.

From Equations (5.28) and (5.32),

$$\omega_o = \frac{1}{RC}, \tag{5.41}$$

and

$$\alpha = \frac{RC + RC + (1 - A_o)\,RC}{RC} = 3 - A_o. \tag{5.42}$$

The natural frequency is determined by the time constant of R and C, the damping coefficient is determined only by A_0, and adjustment is possible without affecting the other parameters of the filter (ω_0 will be unchanged but there will be a small effect on ω_p).

It is unlikely that the value of A_0 required by Equation (5.42) will coincide with the overall gain required. Any required correction can be made by means of an additional, non-frequency-dependent, amplifying stage.

The computer program Equal Component Sallen-Key Filter Design (Appendix A.12) is very similar to Unity Gain Sallen-Key Filter Design (Appendix A.11), but Equations (5.41) and (5.42) have been substituted to allow for the case of equal component values.

5.3.2 High Pass Filter

A second-order high pass filter can be realized by making Z_1 and Z_2 of Figure 5–5 capacitive (C_1 and C_2, respectively) and Z_3 and Z_4 resistive (R_1 and R_2, respectively). Equation (5.23) becomes

$$\frac{V_o}{V_i}(s) = \frac{A_o R_1 R_2}{\frac{1}{sC_1 sC_2} + \frac{R_1}{sC_2} + R_1 R_2 + \frac{R_1}{sC_1} + \frac{R_2(1-A_o)}{sC_1}}. \quad (5.43)$$

Therefore, multiplying through by s^2/R_1R_2,

$$\frac{V_o}{V_i}(s) = \frac{A_o s^2}{s^2 + s\left(\frac{1}{C_2R_2} + \frac{1}{C_1R_2} + \frac{1-A_0}{C_1R_1}\right) + \frac{1}{R_1C_1R_2C_2}}, \quad (5.44)$$

which is of the general form

$$\frac{V_o}{V_i}(s) = \frac{A_o s^2}{s^2 + \alpha\omega_o s + \omega_o^2}, \quad (5.45)$$

where

$$\omega_o = \frac{1}{\sqrt{R_1C_1R_2C_2}} \quad (5.46)$$

as before, and

$$\alpha = \frac{R_1C_1 + R_1C_2 + (1-A_o)\,C_2R_2}{\sqrt{R_1C_1R_2C_2}}, \quad (5.47)$$

which is slightly different from the low pass case (compare Equation [5.32]).

The s^2 (= $-\omega^2$ when $j\omega$ is substituted for s) term in the denominator of Equations (5.44) and (5.45) ensures that the gain increases with frequency. In fact, the response can be shown to be a mirror image of the low pass case, tending to a constant gain at high frequencies as shown in Figures 5–8(a) and 5–9(a). It is important to realize that for all practical operational amplifiers, this "constant" gain will be subject to high frequency roll-off and slew-rate limiting (Sections 2.2 and 2.3).

The frequency of the peak of the response for the lightly damped cases will be higher than ω_0 in this case; the "correction factor" is the reciprocal of that given by Equation (5.38).

The phase response is similar but displaced by 180°. That is, there is a phase lead which tends to 180° at very low frequencies, a lead of 90° at ω_0, and zero phase shift at very high frequencies (neglecting any lag imposed by the operational amplifier itself). These responses are shown in Figures 5–8(b) and 5–9(b).

For the unity gain case, with $R_1 = R_2 = R$, we have

$$\omega_o = \frac{1}{R\sqrt{C_1 C_2}}. \tag{5.48}$$

as before, and

$$\alpha = \frac{C_1 + C_2}{\sqrt{C_1 C_2}} \tag{5.49}$$

For the equal component filter, ω_o and α are given, as before, by Equations (5.41) and (5.42).

5.3.3 Band Pass Filters

A typical band pass filter response is shown in Figure 5–10. The frequency of the peak of the response (the center frequency) is f_o and the higher and lower 3 dB points are f_h and f_l, respectively. Notice that a logarithmic scale is used for frequency and f_0 is the *geometric* mean of f_l and f_h, that is,

$$f_o = \sqrt{f_l f_h}. \tag{5.50}$$

The selectivity of the filter, or "sharpness" of the response, is often specified by the Q (the *quality factor*) of the circuit. This is defined by the ratio of the center frequency and the 3 dB bandwidth, or

$$Q = \frac{f_0}{f_h - f_l}. \tag{5.51}$$

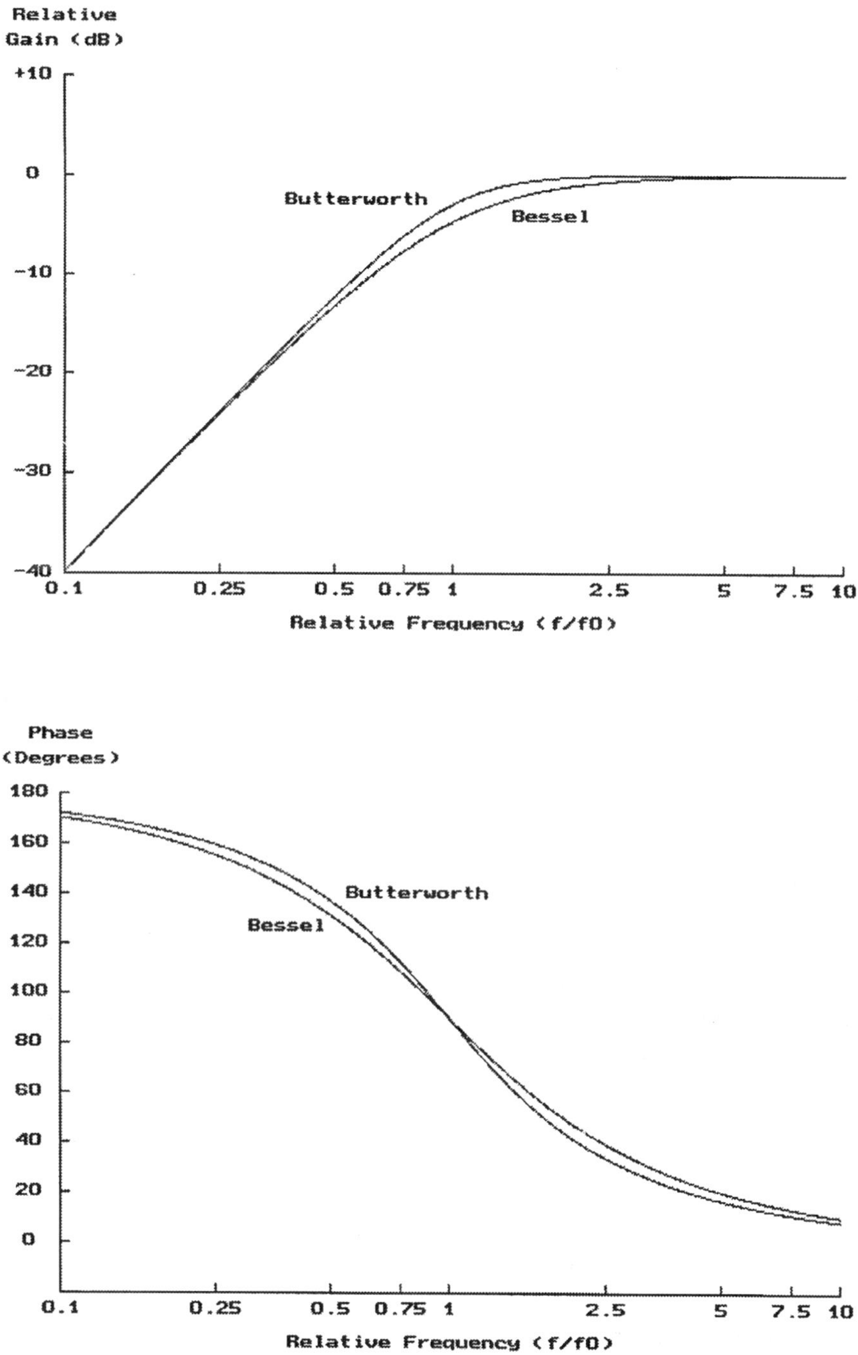

Figure 5–8. Second-order Bessel ($\alpha = 1.732$) and Butterworth ($\alpha = 1.414$) high pass filters: (a) gain (dB) versus frequency (logarithmic scale); (b) phase (degrees) versus frequency (logarithmic scale).

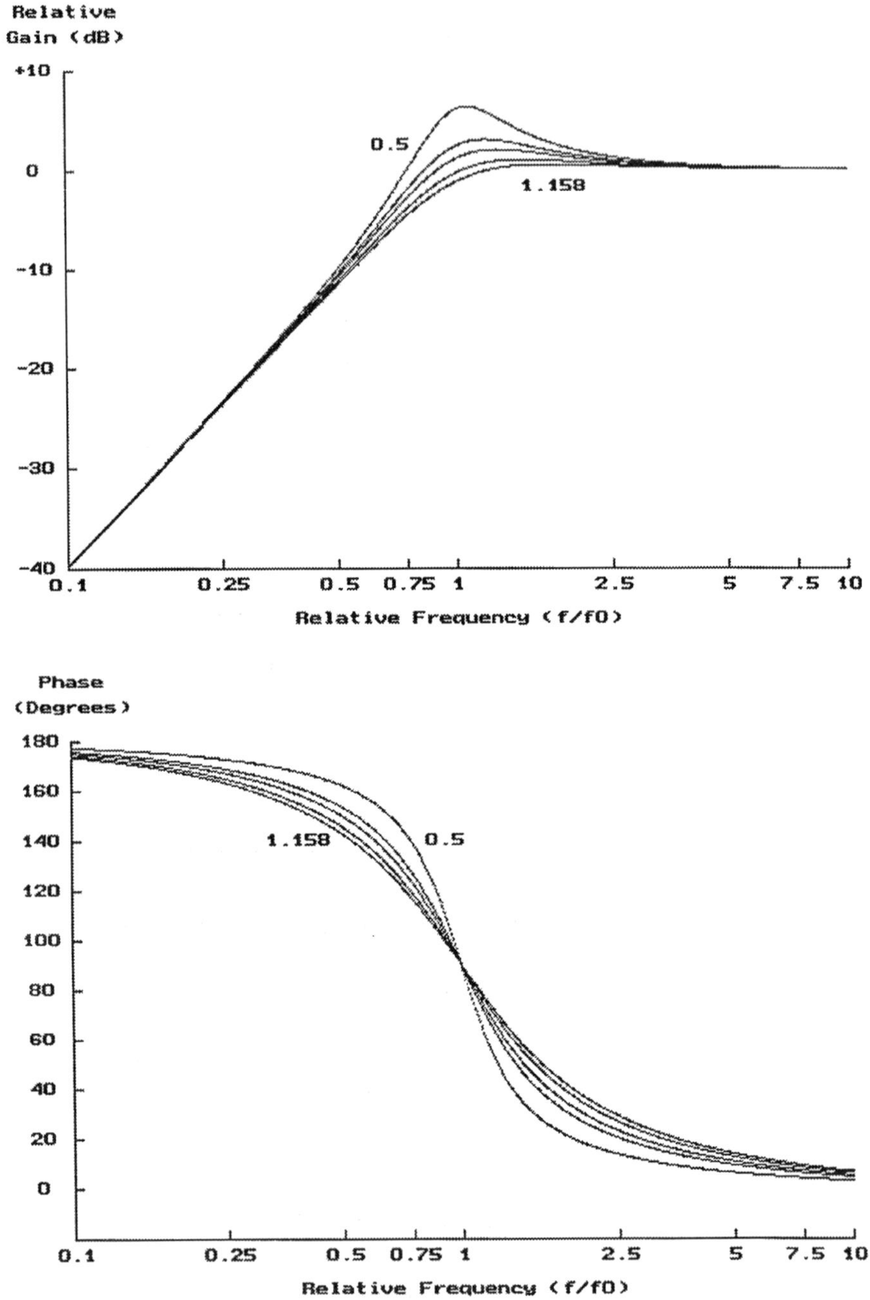

Figure 5-9. Second-order Chebyshev high pass filters for $\alpha = 0.5$, 0.766, 0.866, 1.046, and 1.158; (a) gain (dB) versus frequency (logarithmic scale); (b) phase (degrees) versus frequency (logarithmic scale).

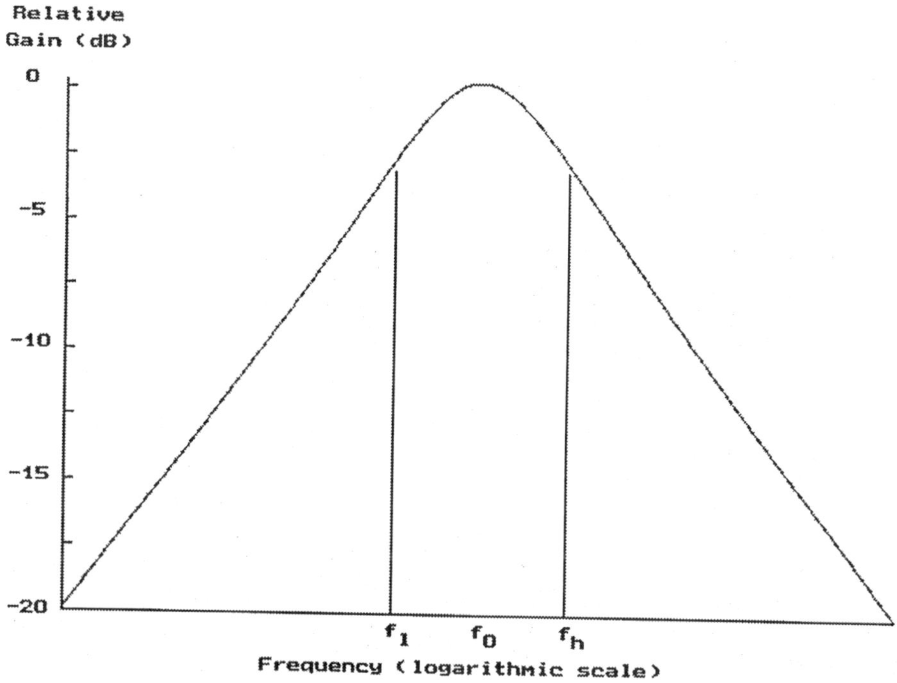

Figure 5–10. Band pass filter response.

A simple band pass filter can be obtained from the general model of Figure 5–5. Z_1 and Z_3 must be made resistive—$Z_1 = R_1$, $Z_3 = R_2$—and Z_2 must be capacitive—$Z_2 = 1/(sC_1)$. For Z_4, a parallel resistance/capacitance combination is required; let this consist of R_3 and C_2, then

$$Z_4 = \frac{\dfrac{R_3}{sC_2}}{R_3 + \dfrac{1}{sC_2}} = \frac{R_3}{1 + sC_2R_3}. \tag{5.52}$$

Substituting these values in Equation (5.23) and rearranging gives

$$\frac{V_o}{V_i}(s) = \frac{\dfrac{A_o s}{C_2R_1}}{s^2 + a\ s + b}, \tag{5.53}$$

where

$$a = \frac{1}{C_1R_1} + \frac{1}{C_1R_2} + \frac{1}{C_2R_1} + \frac{1}{C_2R_3} + \frac{1-A_o}{C_2R_2} \tag{5.54}$$

and

$$b = \left(\frac{1}{C_1C_2R_3}\right)\left(\frac{1}{R_1} + \frac{1}{R_2}\right). \tag{5.55}$$

This unwieldy equation can be simplified by letting $R_1 = R_2 = R$ and $C_1 = C_2 = C$; hence,

$$\frac{V_o}{V_i}(s) = \frac{\dfrac{A_o s}{CR}}{s^2 + \dfrac{s(5-A_o)}{CR} + \dfrac{2}{(CR)^2}}. \tag{5.56}$$

This is of the general form

$$\frac{V_o}{V_i}(s) = \frac{A_o' \alpha\omega_o s}{s^2 + \alpha\omega_o s + \omega_o^2}, \tag{5.57}$$

which is the response of a second-order band pass filter. Notice that A'_o, which is the gain at the center frequency, is not equal to A_o, as was the case for the low- and high-frequency gains of the low and high pass filters, respectively.

The damping coefficient α can be shown to be the reciprocal of the Q of the filter. Therefore, from Equations (5.56) and (5.57) we have

$$\omega_o = \frac{\sqrt{2}}{CR} = 2\pi f_o, \tag{5.58}$$

$$\alpha = \frac{1}{Q} = \frac{5-A_o}{\sqrt{2}}. \tag{5.59}$$

Also,

$$A_o' = \frac{A_o}{5-A_o}. \tag{5.60}$$

Several points should be noted in connection with this filter. The low- and high-frequency roll-off (Figure 5–10) tends to have a slope of 20 dB per decade. However, particularly for high Q values, the slopes can be much greater in the vicinity of f_o (see Figure 5–11). Notice this value of 20 dB per decade for a second-order filter which arises because the characteristic 40 dB per decade of second-order low and high pass filters is effectively "shared" between the rising and falling parts of the characteristic.

Equation (5.59) imposes practical limits on the range of Q values that can be obtained. For high values of Q (A_0 approaching 5), the filter performance becomes extremely sensitive to parameter variations, particularly A_0 itself. For this reason, Q values greater than 10, or at the most 20, are not recommended. For higher Q values, several cascaded stages (Jacob 1993, 420–25) or a state variable filter (Section 5.4) should be used.

Similarly, very low values of Q lead to small and even negative values of A_0. In this (wide bandwidth) case, separate low and high pass filters, as discussed in the previous sections, are to be preferred.

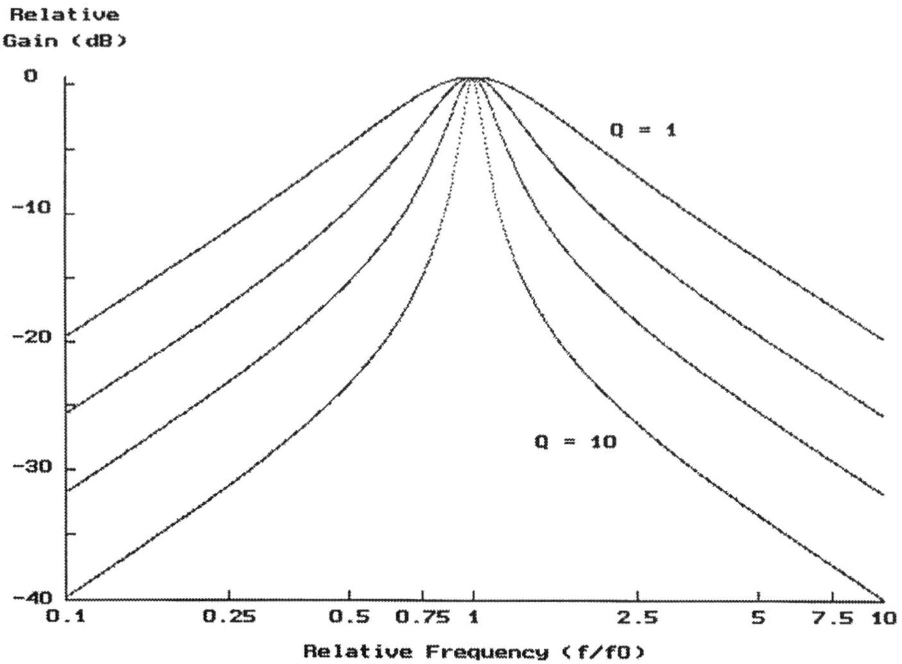

Figure 5–11. Band pass filter responses for $Q = 1, 2, 4$, and 10 .

The computer program Bandpass Filter Design (Appendix A.13) uses Equations (5.58), (5.59), and (5.60) for the design of a simple band pass filter. Either Q or 3 dB bandwidth may be specified.

5.4 State Variable Filters

Originally, the principal application of operational amplifiers was in analog computers. Configurations of operational integrators and summers enabled rapid continuous solutions to be obtained for a wide range of differential equations; the classical text was Korn and Korn (1964). The use of analog computers has now been almost entirely superseded by the use of digital computer simulation techniques.

However, one application of analog computing techniques is still widely used in the form of the so-called *state variable filter*. These can be of any order, but here attention will be devoted to the second-order version (Figure 5–12). This circuit bears a strong resemblance to an analog simulation diagram for a linear, second-order system such as a feedback controller. For an analog simulation diagram see Dorf (1992, 76); for an explanation of state variables, after which the filter is named, see the same work (431–34).

The first stage is a differential amplifier that is complicated slightly by the use of the noninverting input to add V_i and V_{OB}. In Figure 5–12, several groups of components have been given equal values. This is not essential but helps to simplify the analysis. Since it is assumed that no current flows into the

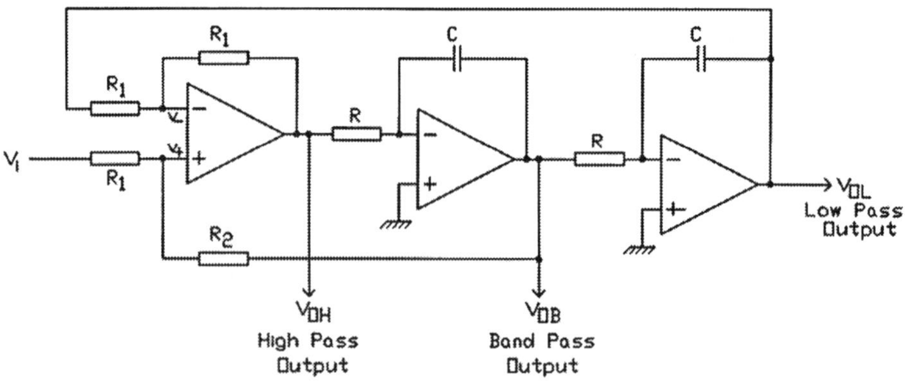

Figure 5–12. Second-order state variable filter.

amplifier, the voltage at the noninverting input of the first amplifier will be V_i plus the difference between V_{OB} and V_i scaled by the potentiometer action of R_1 and R_2, or

$$v_+ = V_i + (V_{OB} - V_i)\frac{R_1}{R_1 + R_2}. \tag{5.61}$$

Hence,

$$\begin{aligned} v_+ &= V_i\left(1 - \frac{R_1}{R_1 + R_2}\right) + V_{OB}\frac{R_1}{R_1 + R_2} \\ &= \frac{V_i R_2 + V_{OB} R_1}{R_1 + R_2}. \end{aligned} \tag{5.62}$$

Summing currents at the inverting input of the first amplifier in Figure 5–12,

$$\frac{V_{OL} - v_-}{R_1} + \frac{V_{OH} - v_-}{R_1} = 0. \tag{5.63}$$

Therefore, since the input and feedback resistors have the same value (R_1),

$$V_{OL} + V_{OH} - 2v_- = 0. \tag{5.64}$$

Noting that, for a high open loop gain, $v_+ \approx v_-$, and substituting Equation (5.62) into (5.64),

$$V_{OL} + V_{OH} - 2\frac{V_i R_2 + V_{OB} R_1}{R_1 + R_2} = 0 \tag{5.65}$$

$$\therefore V_{OH} = 2\frac{V_i R_2 + V_{OB} R_1}{R_1 + R_2} - V_{OL}. \tag{5.66}$$

The second and third stages are simple integrators (Section 4.2). In Laplace form, Equation (4.5) may be written

$$V_o(s) = -\frac{\omega_o}{s}, \tag{5.67}$$

where $\omega_o = 1/(RC)$ and the initial condition has been neglected. Hence, in Figure 5–12

$$V_{OB} = -s\frac{V_{OL}}{\omega_o},$$
$$V_{OH} = s^2\frac{V_{OL}}{\omega_o^2}. \tag{5.68}$$

Substituting so that (5.66) provides V_{OL} as a function of V_i,

$$\frac{V_{OL}}{V_i}(s) = \frac{2\left(\frac{R_2}{R_1+R_2}\right)\omega_o^2}{s^2+2\left(\frac{R_1}{R_1+R_2}\right)\omega_o s+\omega_o^2}, \tag{5.69}$$

which is of exactly the same form as Equation (5.27), where

$$A_0 = 2R_2/(R_1+R_2) \tag{5.70}$$

$$\alpha = 2\zeta = 2R_1/(R_1+R_2) \tag{5.71}$$

and $\omega_o = 1/(RC)$, as it is for the integrators. That is, the relation between V_i and V_{OL} is precisely that of a second-order low pass filter (hence the notation).

The effective output at V_{OB} may be obtained from Equation (5.69) by means of (5.67) as

$$\frac{V_{OB}}{V_i}(s) = -\frac{2\left(\frac{R_2}{R_1+R_2}\right)\omega_o s}{s^2+2\left(\frac{R_1}{R_1+R_2}\right)\omega_o s+\omega_o^2}, \tag{5.72}$$

which is of the same form as Equation (5.57) with

$$A'_0 = R_2/R_1 \tag{5.73}$$

and

$$\alpha = 2\frac{R_1}{R_1+R_2} \tag{5.74}$$

as before. Therefore, output V_{OB} provides a band pass response relative to V_i. Similarly,

$$V_{OH}(s) = \frac{s^2}{\omega_o^2} V_{OL}(s) = \frac{2\left(\frac{R_2}{R_1 + R_2}\right)s^2}{s^2 + 2\left(\frac{R_1}{R_1 + R_2}\right)\omega_o s + \omega_o^2}, \tag{5.75}$$

which is of the same form as Equation (5.45) with

$$A_0 = 2\frac{R_2}{R_1 + R_2} \tag{5.76}$$

and

$$\alpha = 2\frac{R_1}{R_1 + R_2} \tag{5.77}$$

as before; this is a high pass response.

Although a second-order state variable filter requires three amplifiers, it has some advantages. First, it is much less sensitive to parameter variations than the single amplifier configuration. Band pass filters with a Q of 50 or more can readily be realized. Second, the provision of low, high, and band pass functions in a single building block is particularly convenient from a practical point of view.

5.5 Band Rejection Filter

The final class of filter to be discussed is that which rejects, or at least substantially reduces, signals at a particular frequency or within a specified range of frequencies.

One approach (see Jacob 1993, 440–46) is to subtract the output of a suitable band pass filter from the required signal, as shown in Figure 5–13. Since the band pass filter negates V_i, the subtraction is effectively performed by an operational adder. The result is a minimum in the output V_o for those frequencies that provide the maximum output from the band pass filter.

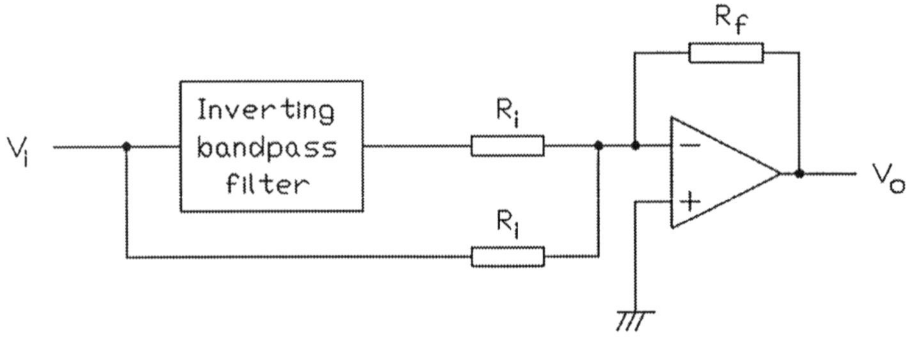

Figure 5–13. Use of a band pass filter to realize a band rejection function.

From Figure 5–13 and Equation (5.57), taking $A'_0 = 1$ for the band pass filter and letting $R_f/R_i = A_o$, it follows that

$$\frac{V_o}{V_i}(s) = -A_o\left(1 - \frac{\alpha\omega_o s}{s^2 + \alpha\omega_o s + \omega_o^2}\right)$$

$$\therefore \frac{V_o}{V_i}(s) = -\frac{A_o(s^2 + \omega_o^2)}{s^2 + \alpha\omega_o s + \omega_o^2}, \qquad (5.78)$$

which is the standard form of a band rejection characteristic.

The twin-T network, as shown in passive form in Figure 5–14, is often used to realize a "notch" characteristic (a narrow bandwidth filter intended for the rejection of one specific frequency). Since there are three capacitors, the transfer function of this network in its general form is of third order. However, provided $R_1C_1 = R_2C_2$ (which generally applies to the component values chosen for a practical realization), it can be shown (see Moschytz 1970, 1105) that cancellation occurs and the transfer function is reduced to second order.

In particular, it is customary to make

$R_1 = R$, $R_2 = kR$, and $R_3 = [k/(1 + k)]R$

$$C_1 = C,\ C_2 = C/k, \text{ and } C_3 = C(1 + 1/k), \qquad (5.79)$$

since this greatly simplifies analysis. It is sometimes called the "potentially symmetric" case, since making $k = 1$ causes the network to become symmetrical. It can be shown (Moschytz 1970) that in this case α (in Equation [5.78])

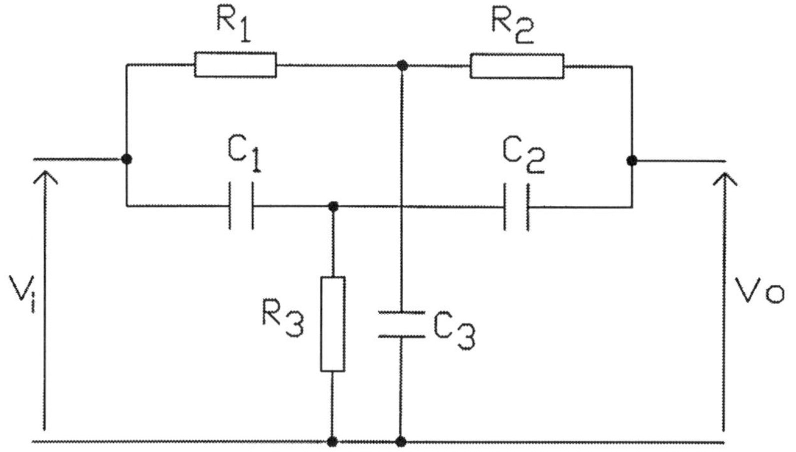

Figure 5–14. Basic twin-T network.

becomes $2(1 + 1/k)$. This means that the maximum value of Q ($= 1/\alpha$) is obtained as k becomes very small and the Q cannot in any case exceed 0.5.

A dramatic increase in Q was obtained following a suggestion by Farrer (1967, 219). This involved connecting the "bottom" of the T to some fraction m of the output, via a buffer amplifier, instead of to ground, as shown in Figure 5–15. The effective value of α now becomes $2(1 + 1/k)(1 - m)$, and large values of Q can be achieved as m approaches 1. Since m now provides adequate control of the effective Q, k can conveniently be made equal to 1 and the filter becomes symmetrical, as shown by the values in Figure 5–15. For these values we have

$$Q = \frac{1}{4(1-m)}, \tag{5.80}$$

$$f_o = \frac{\omega_o}{2\pi} = \frac{1}{2\pi RC}. \tag{5.81}$$

As m tends to 0, the characteristic of the filter approaches that of the passive version (with a buffered output) and the Q tends to its minimum value of 0.25. Larger values of m allow a Q value of around 50 to be readily obtained.

The computer program Notch Filter Design (Appendix A.14) enables suitable component values to be selected.

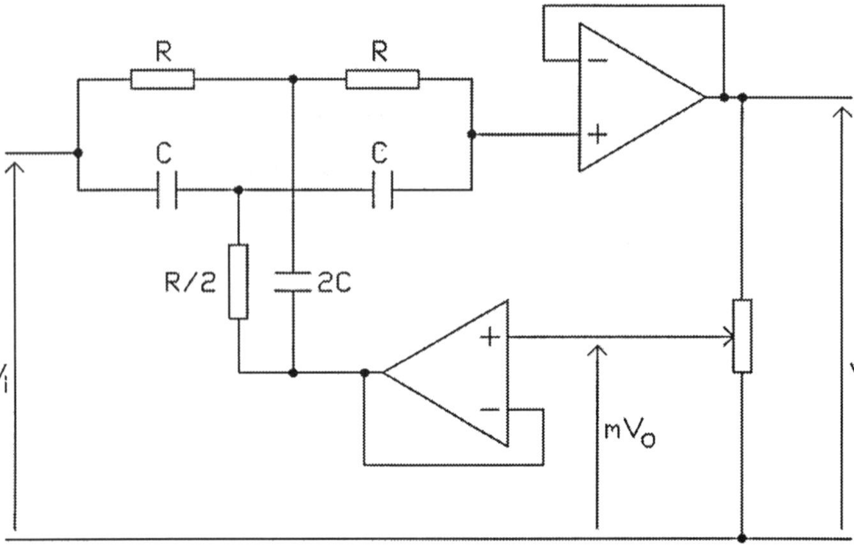

Figure 5–15. Active version of twin-T notch filter.

5.6 References

Brown, J., and E. V. D. Glazier. 1974. *Telecommunications*. London: Chapman and Hall.

Dorf, R. C. 1992. *Modern Control Systems*. 6th ed. Wokingham, England: Addison-Wesley.

Farrer, W. 1967. A simple active filter with independent control over the pole and zero locations. *Electronic Engineering* 39 (470): 219–22.

Jacob, J. M. 1993. *Applications and Design with Analog Integrated Circuits*. 2d ed. Englewood Cliffs, NJ: Prentice Hall International.

Korn, G. A., and T. M. Korn. 1964. *Electronic Analog and Hybrid Computers*. New York: McGraw-Hill.

Moschytz, G. S. 1970. A general approach to twin-T design and its application to hybrid integrated linear active networks. *Bell System Technical Journal* 49 (6): 1105–49.

Sallen, R. P., and E. L. Key. 1955. A practical method of designing RC active filters. *IRE Transactions on Circuit Theory* CT2 (1): 74–85.

CHAPTER 6

Nonlinear Circuits

6.1 Preliminary Comments

With the exception of comparators, only linear applications of operational amplifiers have been discussed in the previous chapters. This means that output magnitude is directly proportional to input magnitude. In the case of multi-input circuits, such as summers, any change in output will still be proportional to the corresponding input change, even in the case where inputs are added with different scaling factors.

In many applications, linear operation is an important requirement; examples are high-quality audio amplifiers and signal analysis systems. The negative feedback inherent in operational amplifier configurations ensures highly linear operation. However, there are applications where deliberately introduced nonlinearity is required. These include rectification and amplitude limiting of signals, waveform shaping (e.g., from triangular to sinusoidal), and dynamic range compression and expansion for communication purposes.

Although a wide variety of nonlinear circuits has been described, two categories can be distinguished:

1. Those based on the use of diodes as switches. In simple cases, such as rectification, a single switching operation may be all that is required. Some examples will be discussed in Sections 6.2, 6.3, and 6.4. More generally, any nonlinear characteristic can be realized as a series of piece-wise linear approximations, using diodes to select each required segment of the characteristic. This approach will be discussed in Section 6.5.

2. Those that exploit the inherent nonlinear characteristic of a particular device. In particular, the bipolar transistor is widely used to generate logarithmic and antilogarithmic functions, as discussed in Section 6.6.

6.2 Simple Limiting

The transfer characteristic of an ideal limiter is shown in Figure 6–1. For input signals between V_{Li-} and V_{Li+}, the output V_o is directly proportional to V_i with a gain that may be selected in the usual way. For inputs exceeding these values, the output remains constant at the appropriate value of V_{Lo-} or V_{Lo+}. Circuits of this kind are particularly useful, for example, in control systems, where it is necessary to protect later parts of the system from overload.

A variety of circuits that can realize, or approximate, this characteristic have been proposed. Two of the more important will now be discussed.

A simple *feedback limiter* is shown in Figure 6–2. For small values of V_i, and hence V_o, both diodes will be reverse-biased, and the circuit will operate as a conventional operational amplifier with a gain of $-(R_f/R_i)$. The negative sign

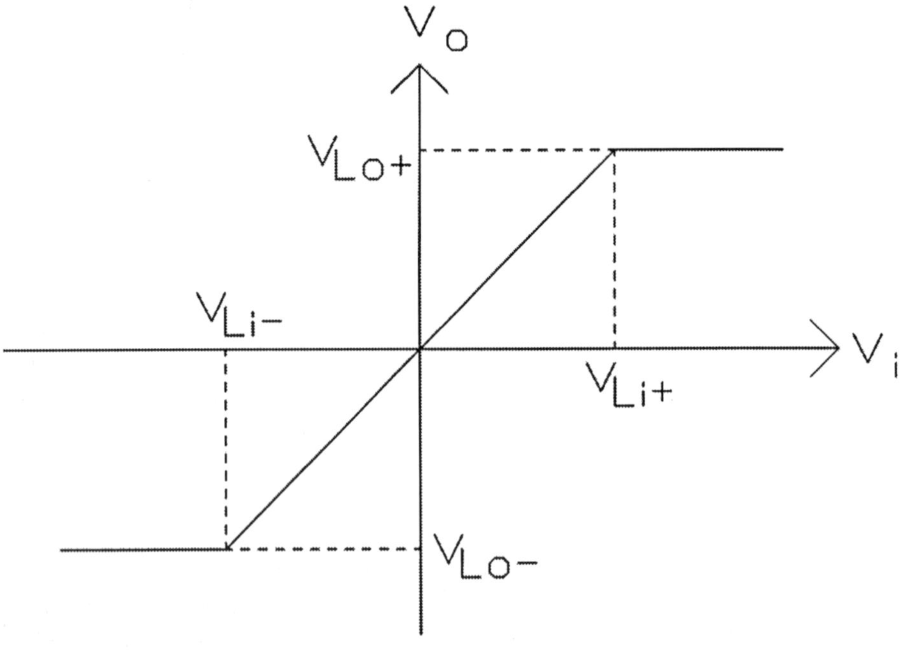

Figure 6–1. Transfer characteristic of ideal limiter.

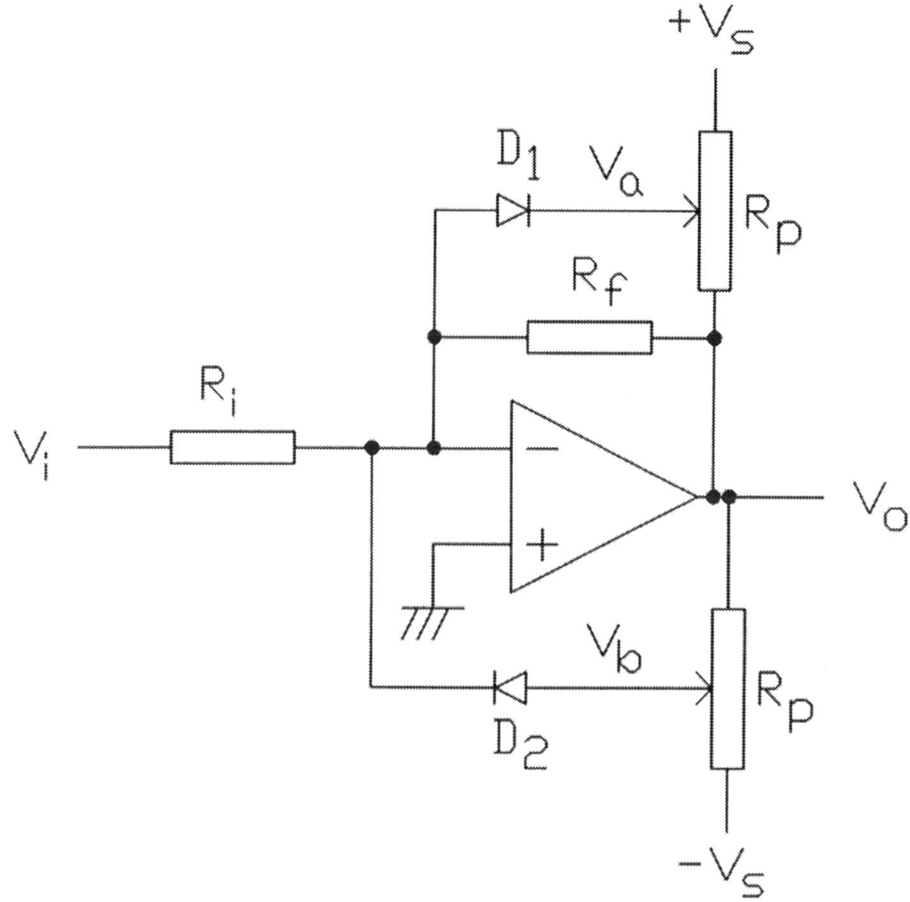

Figure 6–2. Simple feedback limiter.

means that the characteristic of Figure 6–1 will be inverted. The potentiometers R_p connected to $+V_s$ and $-V_s$ represent an additional load on the output of the amplifier that will not otherwise affect its performance. Let aR_p and bR_p represent the portion of R_p between the amplifier output and the appropriate diode. Adjustment of a and b allows the limiting levels to be selected.

As V_i becomes more positive, V_o becomes more negative and the reverse bias V_a on the cathode of diode D_1 will be reduced. Since the anode of this diode is connected to the virtual ground point, it will start to conduct when V_a becomes negative. When this happens, the effective feedback resistance is reduced to aR_p in parallel with R_f. Similarly, when V_i becomes sufficiently negative for D_2 to conduct, the effective feedback resistance becomes bR_p in parallel with R_f.

From Figure 6–2, before the diode conducts,

$$V_a = V_o + a(V_s - V_o), \tag{6.1}$$

since current flow into a reverse-biased diode is negligible. Limiting occurs when $V_a = 0$, or

$$V_o = -V_s \frac{a}{1-a}. \tag{6.2}$$

Since the amplifier gain, before limiting occurs, is $-(R_f/R_i)$, this corresponds to

$$V_{Li+} = V_s \frac{R_i}{R_f}\left(\frac{a}{1-a}\right), \tag{6.3}$$

where V_{Li+} is defined in Figure 6–1.

Similarly,

$$V_{Li-} = -V_s \frac{R_i}{R_f}\left(\frac{b}{1-b}\right). \tag{6.4}$$

The simple feedback limiter, although useful, does not realize the ideal characteristic of Figure 6–1 for several reasons:

1. Practical diodes require an appreciable forward bias (typically 0.5 V) before significant conduction occurs (see Figure 6–3). Equation (6.2) is, therefore, in error by this amount. Normally this effect can be accommodated by a small change in the required values of a and b. However, the forward voltage of semiconductor diodes is temperature-dependent, and this will be reflected in temperature dependence of the limiting levels.

2. Curvature of the diode characteristic (Figure 6–3) means that the discontinuities shown in Figure 6–1 will become rounded. In many applications this is acceptable and, in fact, represents a more accurate simulation of practical limiting effects.

3. Since the effective feedback resistance, when limiting occurs, is not 0, the horizontal portions of the characteristic of Figure 6–1 will actually have a finite slope. However, by making R_p small compared with R_f, the slope can be made acceptably small. For example, if $R_f = 1\ \text{M}\Omega$ and $R_p = 10\ \text{k}\Omega$, the maxi-

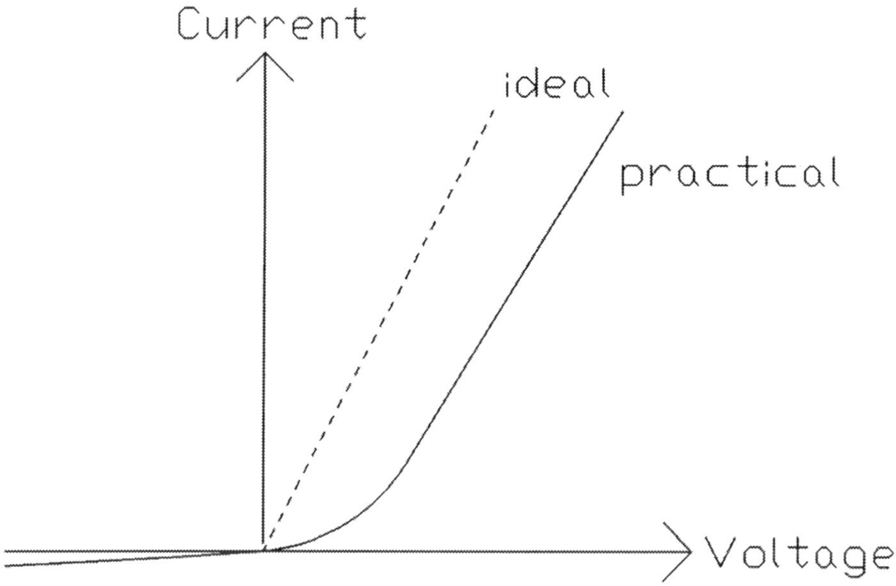

Figure 6–3. Ideal and practical current/voltage characteristics for a diode with series resistance.

mum slope would be approximately 1 in 100, and significantly less for typical values of *a* and *b* that are less than 0.5. For values of *a* and *b* between 0.5 and 1, the diodes can never become forward-biased.

In view of the effects described in (2) and (3) above, the resulting characteristic is sometimes described as "soft" limiting. As before, this closely resembles limiting effects that occur in practice.

The computer program Soft Limiter Design (Appendix A.15) provides assistance in the design of feedback limiters.

6.3 Precision Limiting

For applications where the imperfections of the simple limiter, discussed in the previous section, are not acceptable, performance can be improved by including the switching diode (or diodes) within the amplifier feedback path, as shown in basic form in Figure 6–4.

The nonideal characteristic of the diode (Figure 6–3) may be represented in a general form as a voltage drop V_D that will be some function of the current flowing through it.

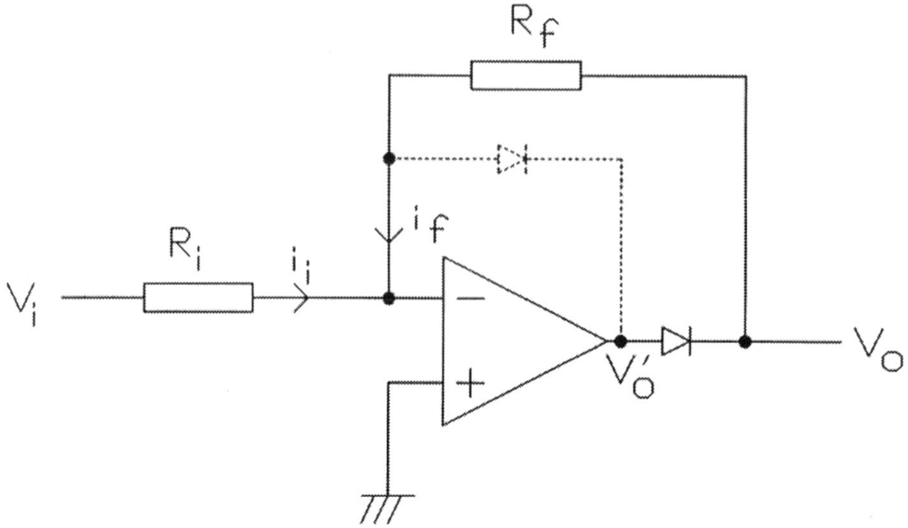

Figure 6–4. Incorporation of diode within the feedback path. The diode shown dotted protects the amplifier from saturation caused by positive inputs.

Analysis of Figure 6–4 in the usual way (Section 1.2) gives

$$\frac{V_i - v}{R_i} + \frac{V_o - v}{R_f} = 0. \tag{6.5}$$

However,

$$V'_o = -Av \tag{6.6}$$

and

$$V_o = V'_o - V_D. \tag{6.7}$$

Therefore,

$$V_o = -Av - V_D \tag{6.8}$$

and

$$v = -\frac{V_o + V_D}{A}. \tag{6.9}$$

Therefore, as A becomes very large, v tends to go to 0 in the usual way, and the error V_D due to the diode is reduced by a factor of A.

This effect is exploited in the precision limiter shown in Figure 6–5. For small inputs, all diodes will be forward-biased and the input of the second amplifier will closely follow the output of the first. Since the diode pairs D_1 and D_2 together with D_3 and D_4 are effectively back to back, errors due to their forward voltages will tend to cancel. Any residual error will be reduced by a factor of A, as explained above.

When the output of the first amplifier becomes sufficiently positive, D_1 will cease to conduct and, in view of the symmetry of the circuit, so will D_4. The output V_o is now determined by $+V_{R1}$ in association with R_1, R_f, and R_L.

Similarly, a negative output from the first amplifier eventually causes D_3 and D_2 to cease conducting. The voltage at the output of amplifier 1 when diodes D_1 and D_4 have just ceased conducting will be determined by $+V_{R1}$ in conjunction with the potential divider formed by R_1 and R_f in parallel with R_L. This voltage will be V_{Lo+} as defined in Figure 6–1. Therefore,

$$V_{Lo+} = V_{R1}\frac{\dfrac{R_f R_L}{R_f + R_L}}{R_1 + \dfrac{R_f R_L}{R_f + R_L}}. \tag{6.10}$$

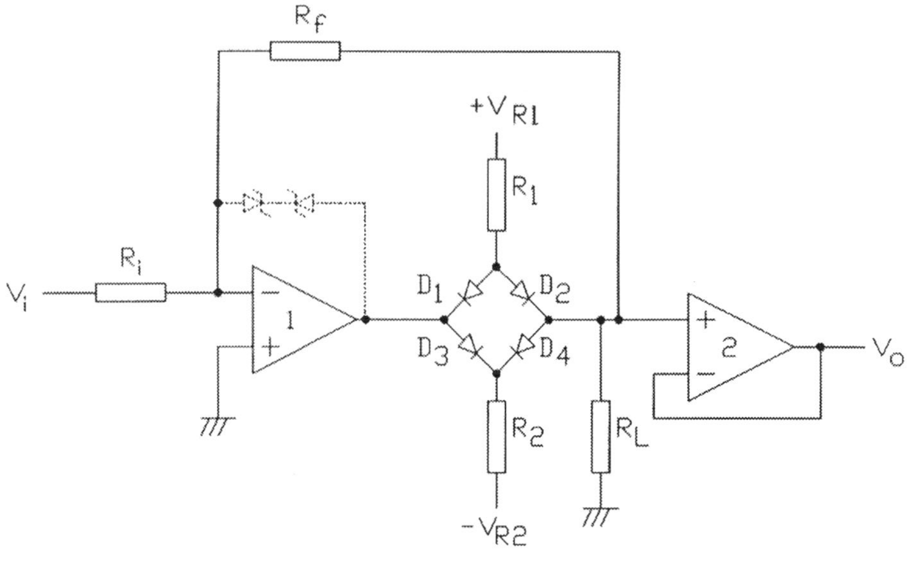

Figure 6–5. Precision limiter circuit using a diode bridge (the Zener diodes shown dotted protect the amplifier from overload).

Hence,

$$V_{Lo+} = V_{R1}\frac{R_f R_L}{R_f R_L + R_f R_1 + R_L R_1}. \tag{6.11}$$

Similarly,

$$V_{Lo-} = V_{R2}\frac{R_f R_L}{R_f R_L + R_f R_2 + R_L R_2}. \tag{6.12}$$

Notice that, when limiting has occurred, the output is independent of V_i and a truly horizontal characteristic is obtained. (The effects of diode leakage current are negligible in view of the current flowing in the other two diodes, which will be forward-biased.) The limiting levels depend on R_L, whose effective value would be changed by the connection of an external load at this point. The buffer amplifier provides the isolation required to avoid this effect. When limiting has occurred, the first amplifier is effectively without feedback. This means that, for large values of V_i, it could be driven heavily into saturation. This will not normally cause damage, but some amplifiers require an appreciable time to recover after the input has been reduced.

For high-speed circuits where this must be avoided, the amplifier can be protected by a feedback limiter of the type discussed in Section 6.2 or, more simply, a pair of Zener diodes, as shown in Figure 6–5. The Zener voltages should be slightly greater than the required limiting values. For each extreme of V_i, one Zener diode will be forward-biased (with a drop of around 0.5 V) and the other will be in its reverse (Zener) region with a voltage drop as specified by the manufacturer for the particular device used. For amplifier outputs that exceed the sum of these two voltages, the effective feedback resistance will become very small and a further increase in the output voltages will be prevented.

The computer program Precision Limiter Design (Appendix A.16) determines the required values of R_1 and R_2 of Figure 6–5 using rearrangements of Equations (6.11) and (6.12).

6.4 Precision Rectification

It is often required to rectify alternating signals with good accuracy for measurement purposes. A common example is the provision of alternating current and voltage ranges on digital multimeters. The practical diode characteristic shown in Figure 6–3 would introduce serious errors, and values less than about 0.5 V could not be measured at all! For this reason, the *precision rectifier circuit* is widely used.

The basic requirements are provided by the circuit of Figure 6–4 where, as already explained, the effects of nonideal diode characteristics are reduced in proportion to the gain of the operational amplifier. This circuit provides precise half-wave rectification with the negative portions of the input providing a positive output, as shown in Figure 6–6. The positive parts of the input are lost and the amplifier can be protected from saturation caused by such inputs by means of the diode shown dotted in Figure 6–4. Negative outputs from the amplifier forward-bias the diode and cause a low effective feedback resistance.

Half-wave rectification is adequate for many applications, but full-wave rectification (Figure 6–7) has the advantage of a higher ripple frequency and, therefore, less stringent filtering requirements. Full-wave rectification involves

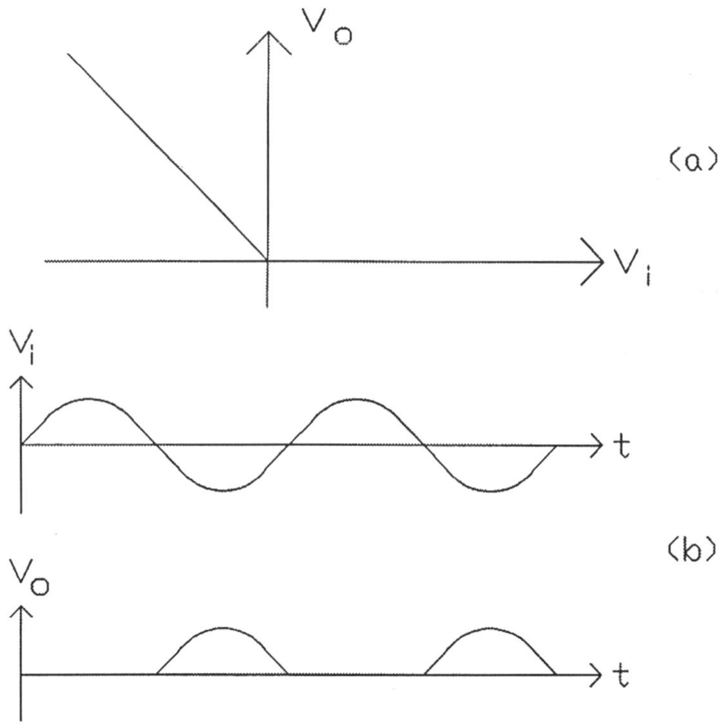

Figure 6–6. Precision half-wave rectifier: (a) transfer characteristic; (b) effect on a sinusoidal input.

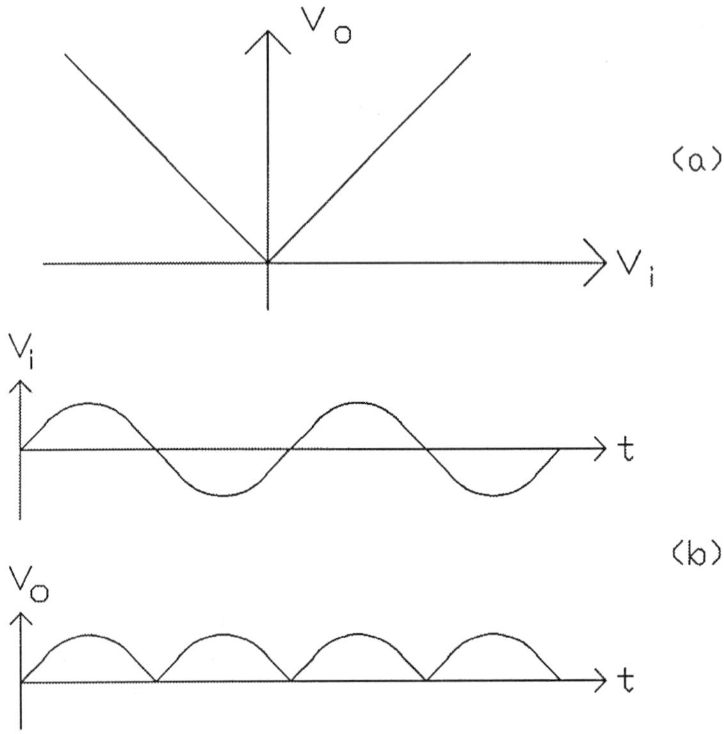

Figure 6–7. Precision full-wave rectifier: (a) transfer characteristic; (b) effect on a sinusoidal input.

a relative sign change for one polarity of input only. The circuit of Figure 6–4 can readily be arranged to provide both positive and negative half cycles, as shown in Figure 6–8. These can then be summed in a conventional differential amplifier (see Section 1.4).

Negative values of V_i provide a positive output from the first amplifier, and feedback is via D_1 and R_{f1}; positive half-cycles appear at the cathode of D_1. Similarly, positive inputs cause feedback via D_2 and R_{f2} with negative half-cycles appearing at the anode of D_2. Notice that, in this configuration, a feedback path is provided for both positive and negative inputs, and additional diodes to avoid saturation are not needed. These half-cycles are then summed differentially in order to provide the required output V_o. R_{f1} and R_{f2} will normally be of equal value and chosen to give the required gain. For simplicity, the differential amplifier has been arranged for unity gain, but this arrangement is not essential.

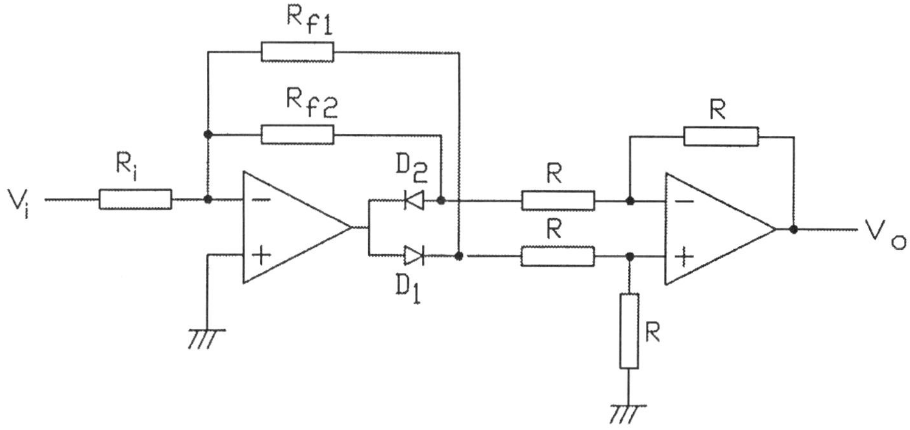

Figure 6–8. Full-wave precision rectifier circuit.

For measurement applications, a ripple-free output from the rectifier is normally needed. This can be achieved by means of low pass filtering. It may also be required to block dc and very low frequency components in the input; this implies high pass filtering. The precision rectifier circuit may therefore be used in conjunction with an active filter configuration (see Section 5.3 and Garrett (1981, 47–49)).

6.5 Arbitrary Function Generators

The circuits discussed in Sections 6.2, 6.3, and 6.4 used diodes in order to realize certain specific simple nonlinear functions. These techniques can readily be extended in order to realize any required function by means of a series of piece-wise linear approximations. Inevitably some error will be involved, but this can be made small by using a sufficiently large number of segments in the approximation. A detailed analysis of the errors involved is beyond the scope of this book; for further details see, for example, Ritchie and Young (1959) and Ream (1959).

One possible approach to diode function generation is shown in basic form in Figure 6–9. Each diode is biased from a negative reference supply (which could be one of the operational amplifier power supplies if this is sufficiently stable) via a resistance R_{Bj} (where j takes a value between 1 and n). The bias arrangement is such that, for negative and small positive values of V_i, all diodes are reverse-biased and V_o is given by $-V_i(R_f/R_i)$ as usual. As V_i becomes

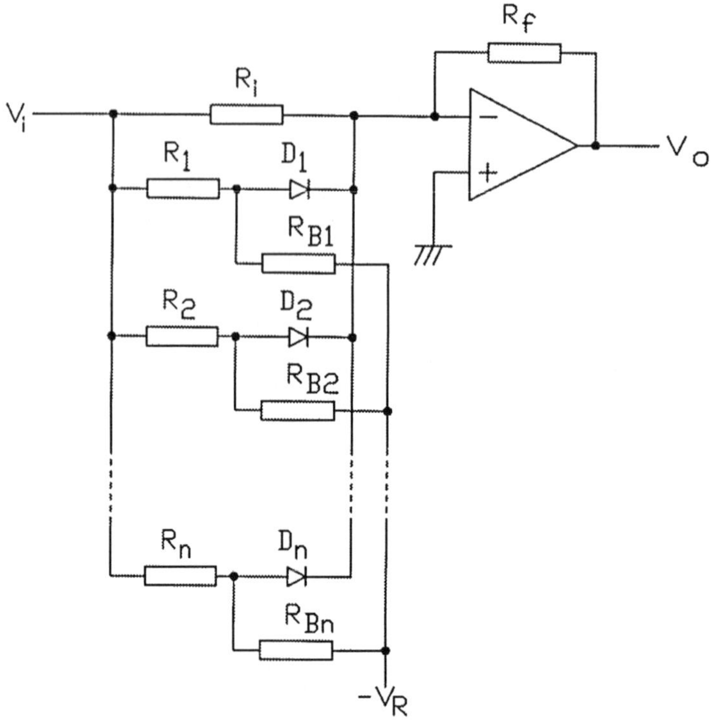

Figure 6–9. Basic diode function generator.

more positive, the diodes will start to conduct. As this happens, the effective input resistance is reduced since R_1, R_2, etc., become effectively connected in parallel with R_i. The type of transfer characteristic that results is shown in Figure 6–10.

Since the cathode of each diode is connected to the virtual ground point, conduction will start for the jth diode D_j when

$$-V_r + \left(\frac{R_{Bj}}{R_j + R_{Bj}}\right)(V_i + V_R) = 0, \tag{6.13}$$

that is,

$$V_i = \frac{V_R R_j}{R_{Bj}}. \tag{6.14}$$

(These values will of course be in error by about 0.5 V due to the forward voltage drop of nonideal diodes.)

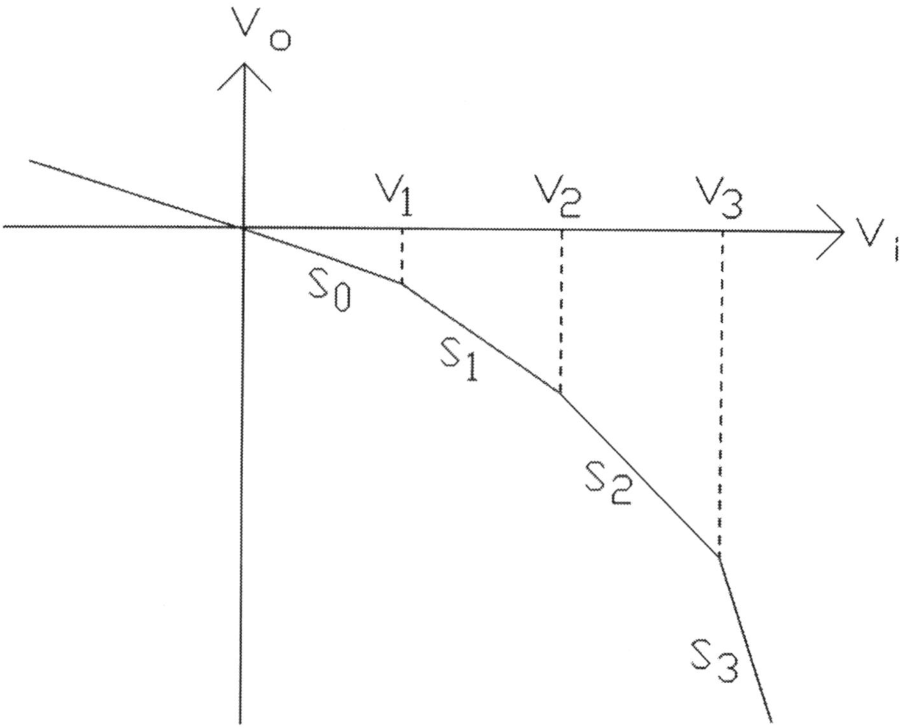

Figure 6–10. Transfer characteristic of basic diode function generator.

Assuming that values are chosen such that D_1 conducts first, followed by D_2 and so on, the slopes shown in Figure 6–10 will be given by

$$S_o = -\frac{R_f}{R_i}, \tag{6.15}$$

$$\begin{aligned} S_1 &= -\frac{R_f}{R_i \text{ and } R_1 \text{ in parallel}} \\ &= -R_f\left(\frac{1}{R_i} + \frac{1}{R_1}\right). \end{aligned} \tag{6.16}$$

In general,

$$S_j = -R_f\left(\frac{1}{R_i} + \frac{1}{R_1} + \frac{1}{R_2} + \cdots + \frac{1}{R_j}\right). \tag{6.17}$$

The circuit described has two major restrictions: 1) it operates for only one polarity of input, and 2) additional input resistors can only increase the slope of the characteristic; that is, $S_1 > S_0$, $S_2 > S_1$, etc. This is a *monotonic* (single-slope) characteristic. The restriction on input polarity can be removed by introducing a second, parallel-connected, input network of the kind shown in Figure 6–9. The polarities of the diodes and the reference voltage should be reversed; the characteristic, however, will still be monotonic with an increasing slope.

Monotonic function generators with a decreasing slope can be realized by connecting a similar diode resistor network in the feedback path of the amplifier (with the diodes connected to the summing junction and the "input resistors" connected to the amplifier output). Subsequent parallel connection of feedback resistances causes a decrease in the slope of the characteristic. Further details are given by Jacob (1993, 498–504).

Monotonic function generators, as described in this section, are capable of providing the characteristics that are normally required such as square law, cube law, and sine wave shaping. More complex circuits have been devised that can generate an arbitrary sequence of slopes; a detailed description has been given by Korn and Korn (1964).

Accurate design of a general diode function generator is complex; the computer program Diode Function Generator Design (Appendix A.17) has been simplified by making several assumptions, as follows:

1. The function to be realized is monotonic with an increasing slope. Decreasing slopes could be accommodated by a network in the feedback path, but this has not been included in the program.

2. The full-scale input and output voltages are equal to each other.

3. The breakpoints are equally spaced with respect to the input voltage range. Improved accuracy can be obtained if more breakpoints are used in regions where the required function changes rapidly (Ritchie and Young 1959, Ream 1959).

4. The diodes are ideal. In practice, the diodes will have a forward voltage drop (about 0.5 V) and a finite "on" resistance. The former could be allowed for by a suitable modification to Equation (6.14) and the latter by ensuring that the input resistances used are large compared with the diode resistances. Notice that if a breakpoint of less than about 0.5 V is required it would, in practice, be necessary to introduce a diode and reference supply of opposite polarity.

5. The required function must be finite over the whole input range. A tangent function, for example, could not be generated over the range 0°–90°.

Within these limitations, the program Diode Function Generator Design (Appendix A.17) gives a good feel for the component values required in a typical function generator.

Calculation of R_i from the feedback resistance and required initial slope is straightforward, but determination of the required input resistances from Equation (6.17) is slightly more difficult. For the second slope (S_1) (ignoring the negative sign),

$$S_1 = R_f\left(\frac{1}{R_i} + \frac{1}{R_1}\right). \tag{6.18}$$

Therefore,

$$R_1 = \left(\frac{S_1}{R_f} - \frac{1}{R_i}\right)^{-1}. \tag{6.19}$$

For each subsequent input resistance an additional term is required, as shown in Equation (6.17), and R_j cannot be calculated until R_{j-1} has been determined. In the program given in Appendix A.17, this is resolved as follows: Let

$$\frac{1}{R_i} + \frac{1}{R_1} + \frac{1}{R_2} + \cdots + \frac{1}{R_j} = K_j \tag{6.20}$$

and

$$K_j = K_{j-1} + \frac{1}{R_j}. \tag{6.21}$$

From (6.17) and (6.20), ignoring the sign again,

$$S_j = R_f K_j. \tag{6.22}$$

Substituting for K_j as given by (6.21),

$$S_j = R_f\left(K_{j-1} + \frac{1}{R_j}\right). \tag{6.23}$$

Therefore,

$$R_j = \left(\frac{S_j}{R_f} - K_{j-1}\right)^{-1}. \tag{6.24}$$

Having obtained the required input resistances, the bias resistances are easily calculated using Equation (6.14).

6.6 Logarithmic Amplifiers

In some special cases, the biased diode networks, discussed in the previous section, can be replaced with a single nonlinear component. In particular, the exponential nature of the voltage/current relationship for a *p-n* semiconductor junction has made simple and effective logarithmic (and antilogarithmic) amplifiers possible.

The logarithmic relationship between amplifier input and output has a wide range of important applications. These include:

1. dynamic range compression and expansion to modify peak to average power ratios in communication systems;

2. generation of products, quotients, and powers by adding, subtracting, and scaling logarithms, respectively; and

3. display of a measured quantity directly in decibels, for example, by a spectrum analyzer.

The required *p-n* junction is conveniently part of a transistor, as shown in Figure 6–11. For an *n-p-n* transistor, the Ebers-Moll equations (Ebers and Moll 1954, 1761–72) give

$$I_C = \alpha I_{ES}\left(e^{\frac{-qV_E}{kT}} - 1\right) - I_{CS}\left(e^{\frac{qV_C}{kT}} - 1\right), \tag{6.25}$$

where

I_C is the collector current,
α is the common base current gain,
I_{ES} and I_{CS} are the emitter and collector saturation currents, respectively,
q is the charge on the electron,
V_E and V_C are the emitter and collector voltages, respectively,

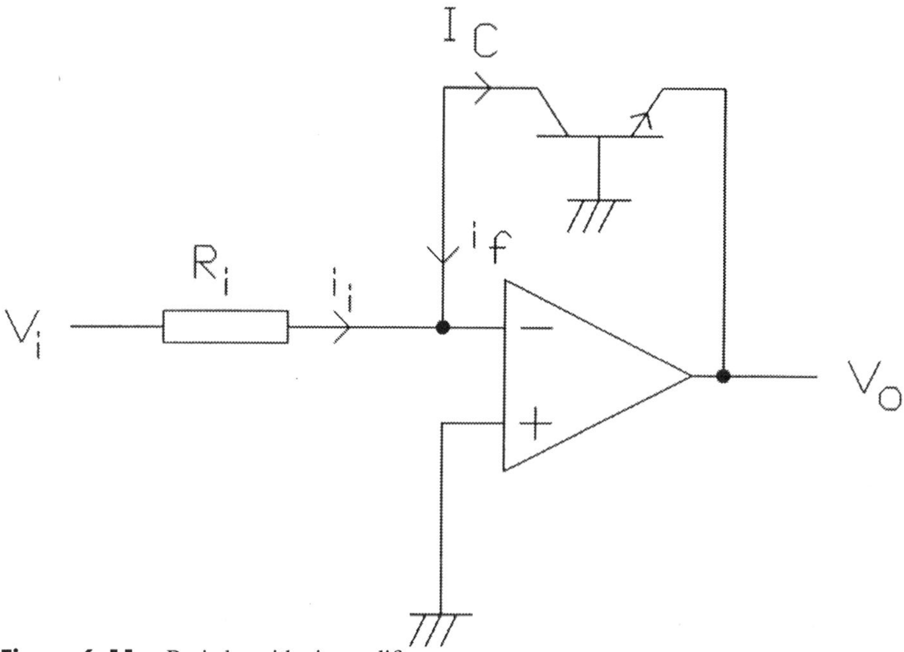

Figure 6–11. Basic logarithmic amplifier.

k is Boltzmann's constant, and
T is the absolute temperature.

In Figure 6–11, the collector is connected to the virtual ground point so, by feedback action, V_C is close to 0, and the second term in Equation (6.25) vanishes. Also, from Figure 6–11, $V_E = V_0$ and $i_f = -I_C$.

Summing currents in the usual way,

$$i_i + i_f = 0, \tag{6.26}$$

$$\frac{V_i}{R_i} = \alpha I_{ES}\left(e^{\frac{-qV_o}{kT}} - 1\right). \tag{6.27}$$

Now I_{ES} is very small (of the order of 10^{-13} A) so, for realistic input and feedback currents, the exponential term will be large compared with 1 and the equation reduces to

$$\frac{V_i}{R_i} = \alpha I_{ES} e^{\frac{-qV_o}{kT}}. \tag{6.28}$$

Taking natural logarithms of both sides,

$$\ln\frac{V_i}{R_i} = -\ln \alpha I_{ES}\left(\frac{qV_o}{kT}\right). \tag{6.29}$$

Therefore,

$$V_o = -\frac{kT}{q}\ln\frac{V_i}{R_i\alpha I_{ES}} = -\frac{kT}{q}\ln\frac{V_i}{V_R}. \tag{6.30}$$

where V_R is a reference voltage equal to $R_i\alpha I_{ES}$.

The output, therefore, is proportional to the logarithm of the input as required. The exponential function necessitated the use of natural logarithms, but base 10 can be accommodated by suitable scaling since $\log_{10}x = \ln x \log_{10}e = 0.4343 \ln x$.

Equation (6.30) shows two points of difficulty. First, the emitter saturation current I_{ES}, although small, varies significantly with temperature and between transistors. Secondly, even if I_{ES} can be compensated, V_o is still directly proportional to temperature.

The first effect is overcome by using a second circuit of the form shown in Figure 6–11. This is fed from a constant (but adjustable) supply V_C. Its output V_{OC}, therefore, from equation (6.30) will be given by

$$V_{OC} = -\frac{kT}{q}\ln\frac{V_C}{R_i\alpha I_{ES}}. \tag{6.31}$$

If the two amplifier outputs are combined in a differential amplifier, the result is

$$V_o' = V_o - V_{OC} = -\frac{kT}{q}\ln\left(\frac{V_i}{R_i\alpha I_{ES}} - \ln\frac{V_C}{R_i\alpha I_{ES}}\right). \tag{6.32}$$

If the same value of R_i is used for each amplifier and the values of α and I_{ES} are similar (ensured by fabrication on the same integrated circuit wafer), this expression reduces to

$$V_o' = -\frac{kT}{q}\ln\frac{V_i}{V_C}, \tag{6.33}$$

and V_C can conveniently be used for calibration.

Unfortunately, the direct dependence of the output on temperature remains. Its effect can be alleviated by means of a noninverting amplifier stage (Figure 1–8) in which the feedback network includes temperature-sensitive elements to provide the required variation of gain with temperature.

It can be seen that it is not easy to design a logarithmic amplifier whose accuracy will be maintained over a wide range of input voltages and temperatures. From a user's point of view the solution is to use a well-established circuit module, of which several are commercially available. These devices are often capable of inverse connection in order to provide an antilogarithm function. A good treatment of amplifiers of this type is provided by Jacob (1993, 518–32).

6.7 References

Ebers, J. J., and J. L. Moll. 1954. The large signal behavior of junction transistors. *Proceedings of the Institute of Radio Engineers* 42: 1761–72.

Garrett, P. H. 1981. *Analog I/O Design*. Reston, VA: Reston Publishing.

Jacob, J. M. 1993. *Applications and Design with Analog Integrated Circuits*. 2nd ed. Englewood Cliffs, N.J.: Prentice Hall International.

Korn, G. A., and T. M. Korn. 1964. *Electronic Analog and Hybrid Computers*. New York: McGraw-Hill.

Ream, N. 1959. Approximation errors in diode function generators. *Journal of Electronics and Control* 7: 83.

Ritchie, C. C., and R. W. Young. 1959. The design of biased diode function generators. *Electronic Engineering* 31: 347.

APPENDIX **A**

Computer Programs

The following programs enable a feel for the magnitudes of the various parameters involved throughout this book. They may also be used, of course, to assist in the design of practical circuits.

Borland Turbo Pascal (version 6.0)[1] for the PC was used, although the programs should run under most versions of Pascal with relatively minor variations. Translation to other languages, particularly C language, should be straightforward for those with some programming experience. Specific CAD packages such as PSpice would also be helpful if available. This approach has been used in the second edition of J. M. Jacob's *Application and Design with Analog Integrated Circuits* (Prentice Hall International, 1993).

No claims are made for elegance or efficiency with respect to these programs. However, they have all been checked with typical values of input data and shown to operate correctly. In many cases, checks are incorporated to protect against input values that can cause program failure, such as a requested gain of 0.5 from a circuit that can only produce gains greater than 1.

Nevertheless, tests of this kind and the associated program testing cannot be exhaustive and there may well be particular combinations of parameter values that cause program failure. Programs may also fail, for example, if a non-numeric character is entered when a numerical value is requested. To keep the programs simple, comprehensive checks for errors of this kind have not been incorporated.

[1]Turbo Pascal® version 6.0 is copyrighted© 1985–1992, by Borland International, Inc., all rights reserved.

A.1 Input and Feedback Resistance Values (Inverting Amplifier)

```
PROGRAM Input_and_Feedback_Resistances;

VAR
   Number_of_inputs, Input_number: integer;
   Vs, Imax, Rl, Rfmin, Rf: real;
   Gain, Ri: array [0..10] of real;
   Ans: char;

BEGIN{Get and check parameter values}
     write('Enter supply voltage: ');
     readln(Vs);
     write('Enter maximum amplifier output current (mA): ');
     readln(Imax);
     write('Enter load to be driven (kohms): ');
     readln(Rl);
     WHILE Imax*Rl <= Vs DO
        BEGIN
             write('Load resistance is too small, try
                    another value: ');
             readln(Rl);
        END;
     write('Enter number of inputs required: ');
     readln(Number_of_inputs);
     FOR Input_number := 1 TO Number_of_inputs DO
         BEGIN
            write('Gain required for input number ',
Input_number, ': ');
            readln(Gain[Input_number]);
         END;

REPEAT      {calculate minimum feedback resistance}
using Equation (1.14)}
     Rfmin := (Rl * Vs) / (Imax * Rl - Vs);
     writeln('The minimum feedback resistance is ',
Rfmin:8:3,' kohms');
     write('Choose a convenient value: ');
     readln(Rf);
     WHILE Rf < Rfmin DO
        BEGIN
```

```
            writeln('Feedback resistance too small, try
                     another value');
            readln(Rf);
        END;
      FOR Input_number := 1 TO Number_of_inputs DO
        BEGIN{Use Equation (1.7)}
          Ri[Input_number] := Rf / Gain[Input_number];
          writeln('Input resistance number ',
Input_number,' = ',
          Ri[Input_number]:8:3,' kohms')
        END;
      writeln('Feedback resistance = ', Rf:8:3,' kohms');
      write('Are any input resistances too low (Y/N)?: ');
      readln(Ans);
UNTIL upcase(Ans) = 'N';
      write('Press <enter> to quit:');
      readln;
END. {Input_and_Feedback_Resistances}
```

A.2 Input and Feedback Resistance Values (Noninverting Amplifier)

```
PROGRAM Noninverting_Input_and_Feedback_Resistances;

VAR
   Number_of_inputs: integer;
   Vs, Imax, Rl, Rmin, Rf, Gain, Ri: real;

BEGIN       {Get and check parameter values}
     write('Enter supply voltage: ');
     readln(Vs);
     write('Enter maximum amplifier output current (mA): ');
     readln(Imax);
     write('Enter load to be driven (kohms): ');
     readln(Rl);
     WHILE Imax*Rl <= Vs DO
        BEGIN
          writeln('Load resistance is too small, try
                   another value');
          readln(Rl);
        END;
     write('Enter required Gain: ');
```

```
     readln(Gain);
     WHILE Gain <= 1 DO
          BEGIN
             write('Gain must be greater than 1, try
                    another value: ');
             readln(Gain);
          END;
                                   {calculate resistance values}
     Rmin := (Rl * Vs) / (Imax * Rl - Vs);  {Equation
        (1.14)}
     writeln('The minimum value of Rf + Ri is ',Rmin:8:3,'
              kohm');
     write('Choose a convenient value for Rf: ');
     readln(Rf);
     WHILE Rf < Rmin DO
        BEGIN
          write('Value of Rf is too small, try another
                 value: ');
          readln(Rf);
        END;
     Ri := Rf / (Gain - 1);  {Equation (1.18)}
     writeln('Input resistance Ri = ',Ri:8:3,' kohm');
     writeln('Feedback resistance Rf = ',Rf:8:3,' kohm');
     write('Press <enter> to quit:');
     readln;
END. {Noninverting_Input_and_Feedback_Resistances}
```

A.3 Common Mode Rejection Ratio (CMRR) Determination

```
PROGRAM Common_Mode_Rejection_Ratio;

USES Crt;

VAR
   GainA, GainB, Difference, Ratio, CMRR: real;
   Ans: char;

BEGIN
     clrscr;
     REPEAT
           writeln;
```

```
          write('Enter channel A gain: ');
          readln(GainA);
          writeln;
          write('Enter channel B gain: ');
          readln(GainB);
          Difference := abs(GainA - GainB);
          IF Difference < 1E-10 THEN
             BEGIN
                  writeln;
                  write('Gains are equal, or very nearly
                         so, ');
                  writeln('CMRR is out of range.');
                  writeln;
             END
          ELSE
              BEGIN     {Use Equation (1.41)}
                   Ratio := (GainA + GainB) / (2 *
Difference);
                   CMRR := 20 * (ln(Ratio) / ln(10));
                   writeln;
                   writeln('CMRR = ', CMRR:6:2, ' db');
                   writeln;
              END;
       write('Press <enter> to continue, Q <enter> to
              quit: ');
       read(Ans);
    UNTIL upcase(Ans) = 'Q';
END. {Common_Mode_Rejection_Ratio}
```

A.4 Slew Rate to Bandwidth Conversion

```
PROGRAM Slew_Rate_to_Bandwidth;
            {Convert bandwidth to slew rate and vice versa}
{Using Equation (2.22)}

VAR
   BW, SR, Peaksignal: real;
   Ans: Char;

PROCEDURE BWtoSR;
        BEGIN
          write('Enter full power bandwidth (kHz): ');
```

```
        readln(BW);
        write('Enter peak signal amplitude (volts): ');
        readln(Peaksignal);
        SR := 2 * Pi * Peaksignal * BW * 1E-3;
        writeln('Slew rate = ', SR:8:3,' volts per
                 microsecond');
      END;

PROCEDURE SRtoBW;
      BEGIN
         write ('Enter slew rate (volts per
                 microsecond): ');
         readln(SR);
         write('Enter peak signal amplitude (volts): ');
         readln(Peaksignal);
         BW := SR * 1E3 / (2 * pi * Peaksignal);
         writeln('Full power bandwidth = ', BW:8:3,'
                 kHz');
      END;

BEGIN; {main program}
     REPEAT
          writeln('To convert bandwidth to slew rate
                 type B');
          writeln('To convert slew rate to bandwidth
                 type S');
          writeln('To quit type Q');
          readln(Ans);
          IF upcase(Ans) = 'B' THEN BWtoSR;
          IF upcase(Ans) = 'S' THEN SRtoBW;
          writeln;
     UNTIL upcase (Ans) = 'Q';
END. {Slew_Rate_Bandwidth}
```

A.5 Offset Error Determination

```
PROGRAM Offset_Error;

USES Crt;

VAR
   Rf, Ri, Vo, Ib, Rc, Id, Error, Ecomp: real;
```

```
   Ans: char;

BEGIN
   REPEAT       {Get parameters}
      ClrScr;
      writeln;
      write ('Enter feedback resistance (kohms): ');
      readln (Rf);
      writeln;
      write ('Enter input resistance (kohms): ');
      readln (Ri);
      writeln;
      write ('Enter op amp offset voltage (mV): ');
      readln (Vo);
      writeln;
      write ('Enter op amp bias current (nA): ');
      readln (Ib);
      writeln;
{Calculate offset error using Equation (3.8)}
           Error := Vo * (1 + Rf / Ri) + Rf * Ib * 1E-03;
      writeln('Total output error = ', Error:6:3,' mV');
      writeln;

            {Compensating resistance from Equation (3.11)}
           Rc := (Rf * Ri) / (Rf + Ri);
      writeln ('Required compensating resistance = ',
                Rc:6:3, ' kohm');
      writeln;
      write ('Enter op amp input offset current (nA): ');
      readln (Id);
      writeln;
           Ecomp := Vo * (1 + Rf / Ri) + Rf * Id * 1E-03;
      writeln ('Compensated output error = ', Ecomp:6:3,'
                mV');
      writeln;
      write('Press <enter> to continue, Q <enter> to
             quit: ');
      read(Ans);
    UNTIL upcase(Ans) = 'Q';
END. {Offset_Error}
```

A.6 Worst Case Drift Performance

```
PROGRAM Drift; {Evaluates worst case drift performance}

USES Crt;

VAR
   Rf, Ri, MaxTemp, VsChange, MaxError, VoltageTempCo: real;
   SupplyRej, BiasTempCo, Voffset, Ib, Vout: real;
   Ans: char;

BEGIN               {Get parameters}
REPEAT
     ClrScr;
     write ('Enter feedback resistance (kohms): ');
     readln (Rf); writeln;
     write ('Enter input resistance (kohms): ');
     readln (Ri); writeln;
     write ('Enter maximum operating temperature (C): ');
     readln (MaxTemp); writeln;
     write ('Enter maximum supply variation (volts): ');
     readln (VsChange); writeln;
     write ('Enter maximum permissible error (mV): ');
     readln (MaxError); writeln;
     write ('Enter offset voltage temperature coefficient
             (microvolts/C): ');
     readln (VoltageTempCo); writeln;
     write ('Enter supply rejection (microvolts/volt): ');
     readln (SupplyRej); writeln;
     write ('Enter bias current temperature coefficient
             (nanoamp/C): ');
     readln (BiasTempCo); writeln;

     {Calculate and check error using Equations (3.12) and
        (3.13)}
     Voffset := (VoltageTempCo * (MaxTemp - 25) * 1E-06)
                  + (SupplyRej * VsChange * 1E-06);
     Ib := BiasTempCo * (MaxTemp - 25) * 1E-09;
     Vout := (Voffset * (1 + Rf/Ri) * 1000) + (Ib * rf *
              1E06);
     writeln ('Worst case error = ', Vout:6:3,' mV');
               writeln;
```

```
    IF Vout <= MaxError THEN
        BEGIN
             writeln ('The amplifier is adequate.');
                       writeln;
             writeln ('Press <enter> to repeat, Q <enter>
                       to quit.');
        END
    ELSE
        BEGIN
            writeln ('The amplifier cannot meet the
                      required specification.');
            writeln;
            writeln ('Press <enter> to try again, Q
                      <enter> to quit: ');
        END;
    read(Ans);
UNTIL upcase(ans) = 'Q';
END. {Drift}
```

A.7 T-Network Feedback

```
PROGRAM T_Network;
{Determines the required component values for a T-network
feedback and compares the resulting offset due to bias
current with that for a single feedback resistance}

USES Crt;

VAR
   Rfeff, Rt, Rs, Rbias, Ib, Vout: real;

BEGIN
     ClrScr;
     write ('Enter required effective feedback resistance
(Mohm): ');
     readln (Rfeff); writeln;
     write ('Choose a value for Rt (kohm): ');
     readln (Rt); writeln;
{Use Equation (3.24)}
     Rs := (Rt * Rt * 1E06) / (Rfeff * 1E06 - (2 * Rt *
        1E03));
     writeln ('Rs = ', Rs:6:3,' ohm'); writeln;
```

```
     Rbias := (Rt * 1E03) + (Rs * Rt * 1E03) / (Rs + (Rt *
        1E03));
     write ('Enter amplifier bias current (nA): ');
     readln (Ib); writeln;
     Vout := Ib * Rbias * 1E-06;
     writeln ('Error due to bias current with T-network =
               ', Vout:6:3,' mV');
     writeln;
     Vout := Ib * Rfeff;
     writeln ('Error due to bias current with ',
               Rfeff:6:3,' megohm');
     writeln ('Feedback resistor = ', Vout:6:3,' mV');
     writeln;
     writeln ('Press <enter> to quit.');
     readln;
END. {T_Network}
```

A.8 Triangular Wave Oscillator Design

```
PROGRAM Triangular_Wave_Osc.;
        {Determines component values for triangular wave
        oscillator}

USES Crt;

VAR
   Rf, Ri, Ratio, Frequency, Period, Timeconstant, C, R:
      real;
   Ans: char;

BEGIN        {Get parameters}
     ClrScr;
     REPEAT
                 writeln ('Rf must be greater than Ri');
                 write ('Enter feedback resistance
                         (kohms): ');
                 readln (Rf);
                 write ('Enter input resistance (kohms): ');
                 readln (Ri);
                 Ratio := Rf/Ri;
     UNTIL  Ratio > 1;
```

```
     write ('Enter required frequency (Hz): ');
     readln (Frequency);
     Period := 1 / Frequency;
     Timeconstant := Ratio * Period / 4;
     writeln ('Required Time Constant is ',
               Timeconstant*1000:8:3, ' msec');

     REPEAT
           write ('Enter integrating capacitor value
                   (microfarad): ');
           readln (C);
           R := Timeconstant / C * 1000;
           writeln ('Required integrator input resistance
                     is ', R:8:3,' kohm');
           writeln ('Is this acceptable (Y/N)?');
           readln (Ans);
     UNTIL upcase(Ans) = 'Y';
END. {Triangular_Wave_Osc}.
```

A.9 Wien Bridge Oscillator Design

```
PROGRAM Wien_Bridge;
        {Determines component values for Wien Bridge
        oscillator}

USES Crt;

VAR
   R, C, Frequency: real;
   Ans: char;

PROCEDURE CalcC;
     BEGIN
       REPEAT
           write ('Enter selected resistance value
                   (kohm): ');
           readln (R);
           C := 1000 / (2 * pi * Frequency * R);
           writeln ('Required capacitance is ', C:8:4, '
                     microfarad');
           writeln ('Try another resistance value (Y/N)?');
           readln (Ans);
```

```
      UNTIL upcase(Ans) = 'N';
    END;

PROCEDURE CalcR;
    BEGIN
      REPEAT
          write ('Enter selected capacitance value
                 (microfarad): ');
          readln (C);
          R := 1000 / (2 * pi * Frequency * C);
          writeln ('Required resistance is ', R:8:4,'
                  kohm');
          writeln ('Try another capacitance value (Y/N)?');
          readln (Ans);
      UNTIL upcase(Ans) = 'N';
    END;

BEGIN {Main Program}
      ClrScr;
      REPEAT
          write ('Enter required frequency (Hz): ');
          readln (Frequency);
          writeln ('Select resistance (R) or capacitance
                  (C)?');
          readln (Ans);
            IF upcase(Ans) = 'R' THEN CalcC;
            IF upcase(Ans) = 'C' THEN CalcR;
            write ('Perform another calculation (Y/N)?');
            readln (Ans);
      UNTIL upcase(ans) = 'N';
END. {Wien_Bridge}
```

A.10 Filter Order Determination

```
PROGRAM FilterOrder;
        {Determines order of filter required for a
specified rejection}

USES Crt;

VAR
  Wantedf, Unwantedf, FrequencyRatio, AttenW, AttenU : real;
```

```
  RequiredAtten, dBperOrder, Attenuation : real;
  Order : integer;
  Ans: char;

BEGIN {Main Program}
      ClrScr;
      REPEAT
            write ('Enter wanted frequency (Hz): ');
            readln (Wantedf);
            write ('Maximum acceptable attenuation at this
                    frequency (dB): ');
            readln (AttenW);
            write ('Enter frequency to be rejected (Hz): ');
            readln (Unwantedf);
            write ('Minimum required attenuation at this
                    frequency (dB): ');
            readln (AttenU);
            FrequencyRatio := Wantedf / Unwantedf;
            IF FrequencyRatio < 1 THEN FrequencyRatio := 1
               / FrequencyRatio;
            RequiredAtten := AttenU - AttenW;
            dBperOrder := 20 * (Ln(FrequencyRatio) /
               Ln(10));
            Order := 0;
            REPEAT
                  Order := Order + 1;
                  Attenuation := Order * dBperOrder;
            UNTIL Attenuation >= RequiredAtten;
            writeln ('A filter of order ', Order, ' is
                       required.');
            writeln;
            write ('Perform another calculation? (Y/N): ');
            readln (Ans);
      UNTIL upcase(ans) = 'N';
END. {FilterOrder}
```

A.11 Unity Gain Sallen-Key Filter Design

```
PROGRAM Sallen_Key_Unity;
 {Determines the component values required for a specified
Sallen-Key unity gain filter}
```

```
VAR  alpha, PeakGain, f3dB, f0, fp, omega0, C1, C2, R :
real;
     Filtertype : integer;

    BEGIN    { Get value of alpha }
      REPEAT
         writeln ('Which type of filter is required,');
         write ('Bessel (1), Butterworth (2) or Chebyshev
                (3)?');
         readln (Filtertype);
        UNTIL (Filtertype = 1) OR (Filtertype = 2) OR
           (Filtertype = 3);
         writeln;
             IF Filtertype = 1 THEN alpha := 1.732;
             IF Filtertype = 2 THEN alpha := 1.414;
             IF Filtertype = 3 THEN
               REPEAT
                  writeln ('What is the required peak
                            gain?');
                  write('0.1, 0.25, 0.5, 1, 2 or 3 dB : ');
                  readln (PeakGain);
                  If PeakGain = 0.1 THEN alpha := 1.303;
                  If PeakGain = 0.25 THEN alpha := 1.236;
                  If PeakGain = 0.5 THEN alpha := 1.158;
                  If PeakGain = 1 THEN alpha := 1.046;
                  If PeakGain = 2 THEN alpha := 0.8862;
                  If PeakGain = 3 THEN alpha := 0.7666;
                UNTIL (PeakGain = 0.1) OR (PeakGain = 0.25)
                   OR
                        (PeakGain = 0.5) OR (PeakGain = 1)
                           OR
                         (PeakGain = 2) OR (PeakGain = 3);

          {Get value of f0}

           IF Filtertype = 1 THEN
           BEGIN
              write ('The attenuation at f0 is
                      approximately 5 dB');
              writeln (' for a Bessel filter,');
              write ('Enter the required frequency for');
              write (' 3 dB attenuation (Hz): ');
              readln (f3dB);
```

```
      f0 := f3dB * 1.2723
   END;

   IF Filtertype = 2 THEN
   BEGIN
     writeln ('The attenuation at f0 is
               approximately 3 dB');
     write ('For a Butterworth filter, enter');
     write (' the required frequency (Hz): ');
     readln (f0);
   END;

   IF Filtertype = 3 THEN
   BEGIN
     write ('Enter the required frequency of');
     write (' the peak of the response (Hz): ');
     readln (fp);
        f0 := fp * sqrt (2 / (2 - (alpha * alpha)));
   END;

       {Calculate component values}

       omega0 := 2 * pi * f0;

    write ('Choose a value for C1 (microfarad): ');
    readln (C1);
    writeln;
    C2 := (alpha * alpha * C1) / 4;
    R := 1000 / (omega0 * sqrt (C1 * C2));
    writeln ('f0 = ', f0:10:3, ' Hz');
    writeln ('alpha = ',alpha:6:3);
    writeln ('C1 = ', C1:8:4, ' microfarad');
    writeln ('C2 = ', C2:8:4, ' microfarad');
    writeln ('R = ', R:8:3, ' kohm');
    readln;
END. {Sallen_Key_Unity}
```

A.12 Equal Component Sallen-Key Filter Design

```
PROGRAM Sallen_Key_Equal;
 {Determines the component values required for a specified
Sallen-Key equal component filter}
```

```
VAR  alpha, A0, PeakGain, f3dB, f0, fp, omega0, C, R : real;
     Filtertype : integer;

    BEGIN    { Get value of alpha }
      REPEAT
        writeln ('Which type of filter is required,');
        write ('Bessel (1), Butterworth (2) or Chebyshev
                (3)? ');
        readln (Filtertype);
      UNTIL (Filtertype = 1) OR (Filtertype = 2) OR
         (Filtertype = 3);
        writeln;
            IF Filtertype = 1 THEN alpha := 1.732;
            IF Filtertype = 2 THEN alpha := 1.414;
            IF Filtertype = 3 THEN
               REPEAT
                 writeln ('What is the required peak
                           gain?');
                 write ('0.1, 0.25, 0.5, 1, 2 or 3 dB : ');
                 readln (PeakGain);
                 If PeakGain = 0.1 THEN alpha := 1.303;
                 If PeakGain = 0.25 THEN alpha := 1.236;
                 If PeakGain = 0.5 THEN alpha := 1.158;
                 If PeakGain = 1 THEN alpha := 1.046;
                 If PeakGain = 2 THEN alpha := 0.8862;
                 If PeakGain = 3 THEN alpha := 0.7666;
               UNTIL (PeakGain = 0.1) OR (PeakGain = 0.25)
                  OR
                        (PeakGain = 0.5) OR (PeakGain = 1)
                           OR
                        (PeakGain = 2) OR (PeakGain = 3);

                        {Get value of f0}

           IF Filtertype = 1 THEN
           BEGIN
             write ('The attenuation at f0 is approximately
                     5 dB');
             writeln (' for a Bessel filter,');
             write ('Enter the required frequency for');
             write (' 3 dB attenuation (Hz): ');
             readln (f3dB);
              f0 := f3dB * 1.2723
```

```
          END;

          IF Filtertype = 2 THEN
          BEGIN
            writeln ('The attenuation at f0 is
                       approximately 3 dB');
            write ('For a Butterworth filter, enter');
            write (' the required frequency (Hz): ');
            readln (f0);
          END;

          IF Filtertype = 3 THEN
          BEGIN
            write ('Enter the reqired frequency of');
            write(' the peak of the response (Hz): ');
            readln (fp);
            f0 := fp * sqrt (2 / (2 - (alpha * alpha)));
          END;

              {Calculate component values}

               omega0 := 2 * pi * f0;

           write ('Choose a value for C (microfarad): ');
           readln (C);
           writeln;
            R := 1000 / (omega0 * C);
            A0 := 3 - alpha;
           writeln ('f0 = ', f0:10:3, ' Hz');
           writeln ('alpha = ', alpha:6:3);
           writeln ('A0 = ', A0:6:3);
           writeln ('C = ', C:8:4, ' microfarad');
           writeln ('R = ', R:8:3, ' kohm');
           readln;

END. {Sallen_Key_Equal}
```

A.13 Bandpass Filter Design

```
PROGRAM Bandpass_Filter;
        {Determines component values required for a second-
order bandpass filter for a specified Q or bandwidth}
```

```
USES Crt;


VAR
   f0, omega0, Q, BW, A0, A1, C, R : real;
   Ans: char;


BEGIN {Main Program}
   ClrScr;
   write ('Enter center frequency (Hz): ');
   readln (f0); writeln;
     REPEAT
       write ('Do you wish to specify Q or 3 dB bandwidth
              (B) ?: ');
       readln (Ans);
     UNTIL (upcase(Ans) = 'Q') OR (upcase(Ans) = 'B');
       IF upcase(Ans) = 'B' THEN
          BEGIN
            writeln;
            write ('Enter required bandwidth (Hz): ');
            readln (BW);
            Q := f0 / BW;
          END;
       IF upcase(Ans) = 'Q' THEN
          BEGIN
            writeln;
            write ('Enter required Q: ');
            readln (Q);
            BW := f0 / Q;
          END;


       IF Q > 20 THEN
             BEGIN
               writeln ('WARNING! This circuit is likely
                         to be unstable');
               writeln ('for Q values greater than 20');
             END;


       IF (Q > 10) AND NOT (Q > 20) THEN
             BEGIN
               writeln ('WARNING! This circuit is very
                         sensitive to');
```

```
            writeln (' parameter variations with high Q
                  values');
          END;
      IF Q < 1 THEN
          BEGIN
            writeln ('WARNING! This circuit is not
                 suitable');
            writeln (' for Q values less than 1');
          END;

        A0 := 5 - (SQRT(2) / Q);
        A1 := A0 / (5 - A0);

          writeln;
          writeln ('Center Frequency = ', f0:8:3, ' Hz');
          writeln;
          writeln ('Q = ', Q:6:3);
          writeln ('Bandwidth = ', BW:8:3, ' Hz');
          writeln;
          writeln ('A0 = ', A0:8:3);
          writeln ('A1 = ', A1:8:3);
          writeln;
          write ('Choose a capacitance value
                (microfarad): ');
          readln (C);
          omega0 := 2 * pi * f0;
          R := 1000 * sqrt(2) / (omega0 * C);
          writeln ('R = ', R:8:3, ' kohm');
          writeln;
     writeln ('Press <enter> to quit.');
     readln;
END. {Bandpass_Filter}
```

A.14 Notch Filter Design

```
PROGRAM Notch_Filter;
        {Determines component values required for a second-
order bandpass filter for a specified Q or bandwidth}

USES Crt;

VAR
```

```
   f0, Q, BW, m, C, R : real;
   Ans: char;

BEGIN {Main Program}
            ClrScr;
            write ('Enter center frequency (Hz): ');
            readln (f0);
            REPEAT
                  writeln ('Do you wish to specify Q');
                  write (' or 3 dB bandwidth (B) ?: ');
                  readln (Ans);
            UNTIL (upcase(Ans) = 'Q') OR (upcase(Ans) =
               'B');
            IF upcase(Ans) = 'B' THEN
               BEGIN
                    write ('Enter required bandwidth
                            (Hz): ');
                    readln (BW);
                    Q := f0 / BW;
               END;
            IF upcase(Ans) = 'Q' THEN
               BEGIN
                    write ('Enter required Q: ');
                    readln (Q);
                    BW := f0 / Q;
               END;
            IF Q > 50 THEN
               writeln ('This circuit is not suitable for
                         such high Q values');
            IF Q < 0.25 THEN
               writeln ('WARNING! This circuit is not
                         suitable for such low Q values');
            writeln ('Center Frequency = ', f0:10:3, ' Hz');
            writeln ('Q = ', Q:6:3);
            writeln ('Bandwidth = ', BW:10:3, ' Hz');
            write ('Choose a capacitance value
                   (microfarad): ');
            readln (C);
            R := 1000 / (2 * pi * f0 * C);
            writeln ('R = ', R:8:3, ' kohm');
            m := 1 - 1 / (4 * Q);
            writeln ('m = ', m:8:3);
```

```
            writeln;
            writeln ('Press <enter> to quit.');
            readln;
END. {Notch_Filter}
```

A.15 Soft Limiter Design

```
PROGRAM Simple_Limiter;
        {Determines component values for a simple feedback
limiter}

USES Crt;

VAR
   Vs, Imax, Rload, Rleff, Rfmin, Rf, Ri, Rpot, a, b, Gain:
      Real;
   Neglimit, Poslimit, Neglimitslope, Poslimitslope: Real;

BEGIN          {Main Program}
     ClrScr;
             {Get parameters}
     write('Enter supply voltage: ');
     readln(Vs);
     write('Enter maximum amplifier output current (mA): ');
     readln(Imax);

     REPEAT
        write('Enter load to be driven (kohms): ');
        readln(Rload);
        write('Enter resistance of potentiometers
               (kohm): ');
        readln (Rpot);
        Rleff := Rload*(Rpot/2)/(Rload + (Rpot/2));
        IF Imax*Rleff <= Vs THEN
           BEGIN
             writeln('Load resistance is too small, try
                      another value');
             writeln('or use higher resistance
                      potentiometers');
           END;
     UNTIL Imax * Rleff >= Vs;
```

```
Rfmin := (Rleff * Vs) / (Imax * Rleff - Vs);
writeln ('The minimum feedback resistance is '
          ,Rfmin:10:3,' kohms');
write ('Choose a convenient value : ');
readln (Rf);
WHILE Rf < Rfmin DO
    BEGIN
        write ('Feedback resistance too small, try
                another value : ');
        readln (Rf);
    END;
write ('Enter gain required in the linear region: ');
readln (Gain);
Ri := Rf / Gain;

REPEAT
      write ('Enter required negative limit (at
              amplifier output): ');
      readln (Neglimit);
      IF (Neglimit > 0) OR (Neglimit < -Vs) THEN
         BEGIN writeln ('Negative limit must be
                         between zero and');
               writeln ('the negative supply voltage');
               write ('Try another value: ');
               readln (Neglimit);
         END;
UNTIL (Neglimit < 0) AND (Neglimit > -Vs);

REPEAT
      write ('Enter required positive limit (at
              amplifier output): ');
      readln (Poslimit);
      IF (Poslimit < 0) OR (Poslimit > Vs) THEN
         BEGIN writeln ('Positive limit must be
                         between zero and');
               writeln ('the positive supply voltage');
               write ('Try another value: ');
               readln (Poslimit);
         END;
UNTIL (Poslimit > 0) AND (Poslimit < Vs);

      a := Abs(Neglimit) / (Vs + Abs(Neglimit));
```

```
            b := Poslimit / (Vs + Poslimit);
            Poslimitslope := (Rf*b*Rpot)/(Ri*(Rf +
               (b*Rpot)));
            Neglimitslope := (Rf*a*Rpot)/(Ri*(Rf +
               (a*Rpot)));
            writeln;
            writeln ('Required input resistance is ',
                     Ri:8:3,' kohm');
            writeln ('The chosen feedback resistance is ',
                     Rf:8:3,' kohm');
            writeln ('a = ', a:8:3);
            writeln ('b = ', b:8:3);
            writeln ('Positive limit slope = ',
                     Poslimitslope:8:3);
            writeln ('Negative limit slope = ',
                     Neglimitslope:8:3);

            writeln;
            write('Press <enter> to quit:');
            readln;
END. {Simple_Limiter}
```

A.16 Precision Limiter Design

```
PROGRAM Precision_Limiter;
      {Determines component values for a hard limiter
using a diode bridge}

USES Crt;

VAR
   Vs, Vref, Imax, Rl, R1, R2, Rleff, Rfmin, Rf, Ri, Gain:
Real;
   Neglimit, Poslimit: Real;

BEGIN  {Main program}
     ClrScr;      {Get parameters}
     write ('Enter reference voltage: ');
     readln (Vref);
     write ('Enter feedback resistance (kohm): ');
     readln (Rf);
     write ('Enter load resistance Rl (kohm): ');
```

```
    readln (Rl);

    REPEAT
       write ('Enter required negative limit (at amplifier
               output): ');
       readln (Neglimit);
       IF (abs(Neglimit) >= Vref) THEN
          BEGIN
            writeln ('Negative limit must be between zero
                      and');
            writeln ('the negative reference voltage');
            write ('Try another value: ');
            readln (Neglimit);
          END;
    UNTIL (Neglimit < 0) AND (Neglimit > -Vref);

    REPEAT
       write ('Enter required positive limit (at amplifier
               output): ');
       readln (Poslimit);
       IF (Poslimit <= 0) OR (Poslimit >= Vref) THEN
          BEGIN
            writeln ('Positive limit must be between zero
                      and');
            writeln ('the positive reference voltage');
            write ('Try another value: ');
            readln (Poslimit);
          END;
    UNTIL (Poslimit > 0) AND (Poslimit < Vref);

  R1 := ((Rf * Rl) / (Rf + Rl)) * ((Vref / Poslimit) - 1);
  R2 := ((Rf * Rl) / (Rf + Rl)) * ((Vref / Abs(Neglimit))
    - 1);
    writeln ('R1 = ', R1:8:3,' kohm');
    writeln ('R2 = ', R2:8:3,' kohm');

    write ('Press <enter> to quit:');
    readln;
END. {Precision_Limiter}
```

A.17 Diode Function Generator Design

```
PROGRAM Func_Gen;
          {Designs a simple diode function generator.  The
required function must be monotonic with an increasing
slope.}

USES Crt;
VAR
   BpIncrement, FSvoltage, Vref, Vinorm, Rf, Ri: real;
   Vin, Vout, Slope, Rin, Rbias, K: array [0..20] of real;
   Bp, NumberofBp, SegNo: integer;

FUNCTION Generatedfunction(x : Real) : Real;
         BEGIN
              Generatedfunction := x * x;
         END; {Change this to change the function
            generated.}

BEGIN  {Main Program}
     ClrScr;
                              {Get parameter values}
     write ('Enter number of breakpoints: ');
     readln (Numberofbp);
     write ('Enter full scale voltage: ');
     readln (FSvoltage);
     write ('Enter reference voltage: ');
     readln (Vref);
     write ('Enter feedback resistance (kohm): ');
     readln (Rf);
     writeln;

     {Calculate breakpoints, separation between breakpoints
(BpIncrement), and associated input and output voltages}
     BpIncrement := FSvoltage / (NumberofBp + 1);
     Vout[0] :- 0;

     FOR Bp := 1 TO (NumberofBp + 1) DO
       BEGIN
         Vin[Bp] := Bp * BpIncrement;
         Vinorm := Vin[Bp] / FSvoltage;   {normalized input
            voltages}
```

```
      Vout[Bp] := FSvoltage * Generatedfunction(Vinorm);
    END;

  {Calculate slopes as shown in Fig. 6.10}
  FOR SegNo := 0 TO NumberofBp DO
     Slope[SegNo] := (Vout[SegNo + 1] - Vout[SegNo]) /
        BpIncrement;

  {Determine input and bias resistances}
  Ri := Rf / Slope[0];    {Equation (6.15)}
  Rin[1] := 1 / ((Slope[1]/Rf) - (1 / Ri)); {Equation
     (6.19)}
  K[1] := (1 / Ri) + (1 / Rin[1]); {Using (6.20)}

     {Use Equations (6.21) and (6.24) to calculate input
resistances and start at segment 2 to avoid recalculation}
  FOR SegNo := 2 TO Numberofbp DO
    BEGIN
      Rin[SegNo] := 1 / ((Slope[SegNo] / Rf) - K[SegNo -
         1]);
      K[SegNo] := K[SegNo - 1] + (1 / Rin[SegNo]);
    END;

    {Use Equation (6.14) to determine bias resistances}
  FOR SegNo := 1 TO Numberofbp DO
      Rbias[SegNo] := Vref * Rin[SegNo] / Vin[SegNo];

  {Display component values and parameters}

    writeln ('Rf = ', Rf:8:3, ' kohm    Ri = ', Ri:8:3,
             ' kohm');
    writeln;

    {Segment table}
    writeln ('Segment':12, 'Slope':12, 'Rin':12,
             'Rbias':12);
    writeln ('0':12, Slope[0]:12:3, 'Ri only':12, 'Not
             Used':12);
  FOR SegNo := 1 to Numberofbp DO
    writeln (SegNo:12, Slope[SegNo]:12:3,
             Rin[SegNo]:12:3, Rbias[SegNo]:12:3);
    writeln;
```

```
      {Breakpoint table}
      writeln ('Breakpoint':12, 'Vin':12, 'Vout':12);
    FOR SegNo := 0 to Numberofbp - 1 DO
      writeln ((SegNo + 1):12, Vin[SegNo+1]:12:3,
               Vout[SegNo+1]:12:3);
      writeln ('Full scale':12, Vin[Numberofbp+1]:12:3,
                  Vout[Numberofbp+1]:12:3);
      readln;
END. {Func_Gen}
```

Index